Sixth edition

APPLETON & LANGE REVIEW OF
ANATOMY

Royce Lee Montgomery, PhD
Professor
Department of Cell and Developmental Biology
School of Medicine
University of North Carolina
Chapel Hill, North Carolina

Kurt Ogden Gilliland, PhD
Department of Cell and Developmental Biology
School of Medicine
University of North Carolina
Chapel Hill, North Carolina

Appleton & Lange Reviews/McGraw-Hill
Medical Publishing Division

New York Chicago San Francisco Lisbon London
Madrid Mexico City Milan New Delhi San Juan Seoul
Singapore Sydney Toronto

The McGraw-Hill Companies

Appleton & Lange Review of Anatomy, Sixth Edition

1 2 3 4 5 6 7 8 9 0 VNH VNH 0 9 8 7 6 5 4 3 2

ISBN: 0-07-137727-1

Notice

Medicine is an ever-changing science. As new research and clinical experience broaden our knowledge, changes in treatment and drug therapy are required. The authors and the publisher of this work have checked with sources believed to be reliable in their efforts to provide information that is complete and generally in accord with the standards accepted at the time of publication. However, in view of the possibility of human error or changes in medical sciences, neither the authors nor the publisher nor any other party who has been involved in the preparation or publication of this work warrants that the information contained herein is in every respect accurate or complete, and they disclaim all responsibility for any errors or omissions or for the results obtained from use of the information contained in this work. Readers are encouraged to confirm the information contained herein with other sources. For example and in particular, readers are advised to check the product information sheet included in the package of each drug they plan to administer to be certain that the information contained in this work is accurate and that changes have not been made in the recommended dose or in the contraindications for administration. This recommendation is of particular importance in connection with new or infrequently used drugs.

This book was set in Palatino by Circle Graphics.
The editors were Catherine W. Johnson and Lester A. Sheinis.
The production supervisor was Lisa Mendez.
The cover designer was Aimée Nordin.
The indexer was Alexandra Nickerson.
Von Hoffmann Graphics, Inc., was printer and binder.

This book was printed on acid-free paper.

Library of Congress Cataloging-in-Publication Data

Montgomery, Royce L.
Appleton & Lange review of anatomy / Royce L. Montgomery, Kurt Ogden Gilliland.—6th ed.
p. ; cm.
Rev. ed. of: Appleton & Lange review of anatomy for the USMLE Step 1 / Royce L. Montgomery, Gerald A. Montgomery. © 1995.
Includes bibliographical references.
ISBN 0-07-137727-1 (alk. paper)
1. Human anatomy—Examinations, questions, etc. 2. Physicians—Licenses—United States—Examinations—Study guides. I. Title: Appleton & Lange review of anatomy. II. Title: Review of anatomy. III. Gilliland, Kurt Ogden. IV. Montgomery, Royce L. Appleton & Lange review of anatomy for the USMLE Step 1. V. Title.
[DNLM: 1. Anatomy—Examination Questions. QS 18.2 M788a 2003]
QM32 .M65 2002
611'.0076—dc21
2002016672

International Edition ISBN 0-07-121248-5

Contents

Introduction v

1. **The Back** 1
   Answers and Explanations 7
2. **The Upper Limb** 11
   Answers and Explanations 23
3. **The Thorax** 31
   Answers and Explanations 41
4. **The Abdomen** 47
   Answers and Explanations 60
5. **The Pelvis and Perineum** 67
   Answers and Explanations 79
6. **The Lower Limb** 87
   Answers and Explanations 100
7. **The Head and Neck** 109
   Answers and Explanations 132

References 141

Index 143

Introduction

If you are planning to prepare for the United States Medical Licensing Examination (USMLE) Step 1, then this book is designed for you. Here, in one package, is a comprehensive review resource with over 600 examination type anatomy multiple-choice questions with referenced explanations of each answer.

This introduction provides specific information on the USMLE Step 1, information on question types, question-answering strategies, and various ways to use this review.

THE UNITED STATES MEDICAL LICENSING EXAMINATION STEP 1

The United States Medical Licensing Examination Step 1 is a one-day computerized examination consisting of approximately 400 questions to test your knowledge in the basic sciences. It contains multiple-choice questions organized within three dimensions. Each dimension is weighted; however, the projected percentages for these dimensions are subject to change from exam to exam. The three dimensions are: (1) System, (2) Process, and (3) Organizational Level. The application materials illustrate the percentage breakout and offer you a detailed content outline to aid you in your review.

Question Formats

The style and presentation of the questions have been fully revised to conform with the United States Medical Licensing Examinations. This will enable you to familiarize yourself with the types of questions to be expected and provide practice in recalling your knowledge in each format. Following the answer to each question, a reference to a particular and easily available text is provided for further reference and reading.

Each of the chapters contains single-best answer multiple choice questions. In some cases, a group of two or three questions may be related to a situational theme. In addition, some questions have illustrative material (e.g., line illustrations of anatomy) that require understanding and interpretation on your part. Moreover, questions may be of three levels of difficulty: rote memory, memory question that requires more understanding of the problem, and a question that requires both understanding and judgment. In view of the fact that the USMLE Step 1 is moving toward the judgment, critical-thinking type question, we have attempted to write this review with this emphasis.

One Best-Answer-Single Item Question. The majority of the questions are posed in the A type, or "one best answer single item" format. This is the most popular question format in most exams. It generally contains a brief statement, followed by five options of which only ONE is entirely correct. The options on the USMLE are lettered A, B, C, D, and E. Although the format for this question type is straightforward, the questions can be difficult because some of the distractors may be partially right. The instructions you will see for this type of question will generally appear as below:

DIRECTIONS (Question 1): Each of the numbered items or incomplete statements in this section is followed by answers or by completions of the statement. Select the ONE lettered answer or completion that is BEST in each case.

An example of this question type is:

1. An obese 21-year-old woman complains of increased growth of coarse hair on her lip, chin,

chest, and abdomen. She also notes menstrual irregularity with periods of amenorrhea. The most likely cause is

(A) polycystic ovary disease
(B) an ovarian tumor
(C) an adrenal tumor
(D) Cushing's disease
(E) familial hirsutism

In the question above, the key word is "most." Although ovarian tumors, adrenal tumors, and Cushing's disease are causes of hirsutism (described in the stem of the question), polycystic ovary disease is a much more common cause. Familial hirsutism is not associated with the menstrual irregularities mentioned. Thus, the most likely cause of the manifestations described can only be "(A) polycystic ovary disease."

STRATEGIES FOR ANSWERING ONE BEST ANSWER-SINGLE ITEM QUESTIONS

1. Remember that only one choice can be the correct answer.
2. Read the question carefully to be sure that you understand what is being asked. Pay attention to key words like "most."
3. Quickly read each choice for familiarity. (This important step is often not done by test takers.)
4. Go back and consider each choice individually.
5. If a choice is partially correct, tentatively consider it to be incorrect. (This step will help you eliminate choices and increase your odds of choosing the correct answer.)
6. Consider the remaining choices and select the one you think is the answer. At this point, you may want to quickly scan the stem to be sure you understand the question and your answer.
7. If you do not know the answer, make an educated guess. Your score is based on the number of correct answers, not the number you get incorrect. **Do not leave any questions unanswered.**
8. The actual examination is timed for an average of 60 seconds per question. It is important to be thorough to understand the question, but it is equally important for you to keep moving.

Answers, Explanations, and References

In each of the sections of this book, the question sections are followed by a section containing the answers, explanations, and references to the questions. This section (1) tells you the answer to each question; (2) gives you an explanation/review of why the answer is correct and background information on the subject matter; and (3) tells you where you can find more in-depth information on the subject matter in other books and/or journals. We encourage you to use this section as a basis for further study and understanding.

If you choose the correct answer to a question, you can then read the explanation (1) for reinforcement and (2) to add to your knowledge about the subject matter (remember that the explanations usually tell not only why the answer is correct, but also why the other choices are incorrect). **If you choose the wrong answer** to a question, you can read the explanation for a learning/reviewing discussion of the material in the question. Furthermore, you can note the reference cited (e.g., "Joklik et al, pp 103–114"), look up the full source in the bibliography at the end of the section (e.g., "Joklik WK, Willett HP, Amos DB. *Zinsser's Microbiology.* 20th ed. Norwalk, Conn: Appleton & Lange; 1992"), and refer to the pages cited for a more in-depth discussion.

SPECIFIC INFORMATION ON THE STEP 1 EXAMINATION

The official source of all information with respect to the United States Medical Licensing Examination Step 1 is the National Board of Medical Examiners (NBME), 3930 Chestnut Street, Philadelphia, PA 19104. Established in 1915, the NBME is a voluntary, nonprofit, independent organization whose sole function is the design, implementation, distribution, and processing of a vast bank of question items, certifying examinations, and evaluative services in the professional medical field.

Please contact the NBME or visit the USMLE web site (*www.usmle.org*) for information on exam registration and scoring.

CHAPTER 1

The Back

Questions

DIRECTIONS (Questions 1 through 40): Each of the numbered items or incomplete statements in this section is followed by answers or by completions of the statements. Select the ONE lettered answer or completion that is BEST in each case.

1. The vertebral column does all of the following EXCEPT

    (A) protect the spinal cord and spinal nerves
    (B) support the weight of the body
    (C) provide a pivot for the head
    (D) play an important role in posture and locomotion
    (E) form the main part of the appendicular skeleton

2. Which of the following is NOT a distinctive characteristic of a typical cervical vertebra?

    (A) The body is small and wider from side to side than anteroposteriorly.
    (B) The vertebral foramen is large and triangular.
    (C) The transverse processes contain transverse foramina.
    (D) The articular processes contain superior facets directed inferoanteriorly.
    (E) The spinous processes are short and bifid.

3. Which of the following is NOT a distinctive characteristic of a typical thoracic vertebra?

    (A) The body is heart-shaped.
    (B) The vertebral foramen is triangular and larger than in cervical and lumbar vertebrae.
    (C) The transverse processes are long and strong and extend posterolaterally.
    (D) The articular processes contain superior facets directed posteriorly and slightly laterally.
    (E) The spinous processes are long and slope posteroinferiorly.

4. Which of the following is a distinctive characteristic of a typical lumbar vertebra?

    (A) The body is massive and kidney-shaped when viewed superiorly.
    (B) The vertebral foramen is circular and smaller than those of cervical and lumbar vertebrae.
    (C) The transverse processes are long and slender and contain mammillary processes.
    (D) The articular processes contain accessory processes.
    (E) The spinous processes are long and slope posteroinferiorly.

5. Which of the following is true regarding the intervertebral disc between the C1 and C2 vertebrae?

    (A) Its annulus fibrosus is composed of concentric lamellae of fibrocartilage.
    (B) It does not contain a nucleus pulposus as other intervertebral discs do.
    (C) It is thicker than other intervertebral discs.
    (D) It acts like a shock absorber in response to axial forces.
    (E) There is no intervertebral disc between the C1 and C2 vertebrae.

6. All of the following are true regarding the posterior longitudinal ligament EXCEPT
   (A) It is narrower and weaker than the anterior longitudinal ligament.
   (B) It runs within the vertebral canal and connects the vertebral bodies to each other.
   (C) It is provided with pain nerve endings.
   (D) It helps prevent hyperextension of the vertebral column.
   (E) It is attached to the intervertebral discs and the posterior edges of the vertebral bodies.

7. Spinal arteries supplying the vertebrae are branches of the
   (A) vertebral and ascending cervical arteries in the neck
   (B) posterior intercostal arteries in the lumbar region
   (C) subcostal and lumbar arteries in the pelvis
   (D) iliolumbar and lateral and medial sacral arteries in the thorax
   (E) aorta

8. Which of the following back muscles is innervated by dorsal rami?
   (A) latissimus dorsi
   (B) levator scapulae
   (C) rhomboid major
   (D) rhomboid minor
   (E) longissimus

9. Which of the following is NOT a deep (or intrinsic) muscle of the back?
   (A) serratus posterior inferior
   (B) levatores costarum
   (C) iliocostalis
   (D) multifidus
   (E) splenius capitis

10. The splenius capitis and cervicis
    (A) extend the head and neck
    (B) flex the head and neck
    (C) elevate ribs, assisting inspiration
    (D) stabilize the atlas and axis
    (E) move the upper limb

11. The erector spinae muscles
    (A) flex the vertebral column
    (B) flex the head
    (C) control flexion of the back
    (D) prevent lateral bending of the vertebral column
    (E) assist with elevation

12. Which of the following is not a member of the minor deep layer of the back?
    (A) levatores costarum
    (B) cervical intertransversarii
    (C) spinalis
    (D) interspinales
    (E) thoracic intertransversarii

13. The transversospinalis muscles do all of the following EXCEPT
    (A) extend the head
    (B) extend the thoracic and cervical regions of the vertebral column
    (C) elevate ribs, assisting inspiration
    (D) stabilize vertebrae
    (E) assist with rotation of the vertebral column

14. Which of the following muscles does NOT attach to transverse processes of vertebrae?
    (A) semispinalis
    (B) multifidus
    (C) rotatores
    (D) intertransversarii
    (E) spinalis

15. Which of the following muscles does NOT laterally bend the cervical intervertebral joints?
    (A) longus colli
    (B) iliocostalis cervicis
    (C) longissimus capitis and cervicis
    (D) splenius capitis
    (E) splenius cervicis

16. Which of the following muscles are likely to be organs of proprioception instead of producers of motion?

(A) iliocostalis
(B) spinalis
(C) rotatores
(D) splenius capitis
(E) longissimus

17. Which of the following statements is true regarding the suboccipital and deep neck muscles?

(A) The rectus capitis posterior major arises from the spinous process of C2 and inserts into the lateral part of the inferior nuchal line.
(B) The obliquus capitis inferior arises from the posterior tubercle of the posterior arch of C1 and inserts into the medial part of the inferior nuchal line.
(C) The obliquus capitis superior arises from the spinous process of C2 and inserts into the transverse process of C1.
(D) The rectus capitis posterior minor arises from the transverse process of C1 and inserts into the occipital bone between the nuchal lines.
(E) The suboccipital muscles are innervated by the ventral rami of C1.

18. ALL of the following extend the atlanto-occipital joint EXCEPT

(A) rectus capitis posterior major and minor
(B) semispinalis capitis
(C) longus capitis
(D) splenius capitis
(E) longissimus capitis

19. The suboccipital triangle is composed of

(A) a superolateral and superomedial border (the superior oblique and rectus capitis posterior major)
(B) an inferolateral border (semispinalis capitis)
(C) a floor (C2)
(D) a roof (trapezius)
(E) greater and lesser occipital nerves

20. ALL of the following are innervated by dorsal rami EXCEPT

(A) muscles of the suboccipital triangle
(B) skin over the neck and occipital bone (innervated by greater occipital nerve)
(C) intrinsic muscles of the back
(D) skin of the neck and scalp (innervated by lesser occipital nerve)
(E) skin of the central part of the back

21. Which of the following statements is correct?

(A) The spinal cord is enlarged in the thoracic region for innervation of the upper limbs.
(B) The spinal cord is enlarged in the lumbosacral region for innervation of the lower limbs.
(C) In embryos, the spinal cord occupies only the superior two-thirds of the vertebral canal.
(D) In adults, the spinal cord occupies the full length of the vertebral canal.
(E) The cauda equina is composed of ventral but not dorsal roots.

22. Which of the following is NOT true in adults?

(A) The tapering end of the spinal cord may terminate as high as T12 or as low as L3.
(B) The first cervical nerves lack dorsal roots in 50% of people.
(C) The coccygeal nerve may be absent.
(D) The terminal filum is the vestigial remnant of the caudal part of the spinal cord that was in the tail of the embryo.
(E) The spinal cord has a lumbar enlargement for the lower limb but no equivalent enlargement for the smaller upper limb.

23. Which of the following is contained in the extradural (epidural) space?

(A) fat (loose connective tissue)
(B) external vertebral venous plexus
(C) CSF
(D) denticulate ligaments
(E) radicular, medullary, and spinal arteries

24. All of the following are contained in the subarachnoid (leptomeningeal) space EXCEPT

(A) CSF
(B) arachnoid trabeculae
(C) segmental medullary arteries
(D) spinal arteries
(E) internal vertebral plexus

25. Which of the following does NOT give rise to arteries supplying the spinal cord?

(A) ascending cervical artery
(B) deep cervical artery
(C) intercostal arteries
(D) lumbar arteries
(E) thoracoacromial artery

26. Which of the following is true?

(A) There are paired anterior spinal arteries.
(B) There are paired posterior spinal arteries.
(C) The sulcal (central) arteries are formed by the union of branches of the vertebral arteries.
(D) Each anterior spinal artery is a branch of either the posteroinferior cerebellar artery or the vertebral artery.
(E) There is usually one anterior and one posterior spinal vein

27. Which of the following is true?

(A) There are usually two anterior and two posterior spinal veins.
(B) Veins of the spinal cord are distributed in a similar fashion to that of spinal arteries.
(C) Spinal veins are unique in that they do not communicate with each other.
(D) Spinal veins are arranged laterally.
(E) Spinal veins are drained by sulcal and meningeal vein.

28. All of the following are contained in typical spinal nerves EXCEPT

(A) sensory fibers from tendons and joints
(B) motor fibers to muscles
(C) parasympathetic fibers to glands
(D) sensory fibers from blood vessels and glands
(E) motor fibers to smooth muscle

29. Which of the following is not true regarding the parasympathetic nervous system?

(A) The cell body of the presynaptic neuron is located in the gray matter of the CNS.
(B) The cell body of the postsynaptic neuron is located in an autonomic ganglion outside the CNS.
(C) The postsynaptic neuron emits norepinephrine.
(D) Its neurons are craniosacral in origin.
(E) It promotes quiet and orderly processes of the body.

30. Postsynaptic sympathetic fibers that ultimately innervate the body wall and limbs do which of the following?

(A) pass from the sympathetic trunks to adjacent ventral rami through gray rami communicantes
(B) pass from the sympathetic trunks to adjacent ventral rami through white rami communicantes
(C) pass from the sympathetic trunks to adjacent dorsal rami through gray rami communicantes
(D) pass from the sympathetic trunks to adjacent dorsal rami through white rami communicantes
(E) pass from the sympathetic trunks to splanchnic nerves

31. Postsynaptic sympathetic fibers do ALL of the following EXCEPT

(A) constrict the pupil of the eye
(B) stimulate contraction of blood vessels
(C) stimulate contraction of erector pili muscles
(D) cause sudomotion
(E) cause goose bumps

32. Which of the following is NOT true?

(A) Variations in vertebrae are affected by race, sex, genetic factors, and environmental factors.
(B) An increased number of vertebrae occurs more often in males, and a reduced number occurs more often in females.
(C) The number of cervical vertebrae can be 6, 7, or 8.
(D) Some people have more than five lumbar vertebrae and therefore fewer thoracic vertebrae.
(E) The sacrum is typically composed of five fused vertebrae.

33. Which of the following statements about kyphosis is true?

(A) Kyphosis may result from developmental anomalies as well as from osteoporosis.
(B) The vertebral column curves anteriorly.
(C) Kyphosis results in an increase in the lateral diameter of the thorax.
(D) Women may develop a temporary kyphosis during pregnancy.
(E) It is also known as "swayback" or "hollow back."

34. Lordosis is characterized by which of the following?

(A) an abnormal increase in thoracic curvature
(B) an anterior rotation of the pelvis
(C) an abnormal lateral curvature
(D) rotation of the vertebrae
(E) lateral curvature of the spine

35. Scoliosis may be caused by which of the following?

(A) asymmetrical weakness of intrinsic back muscles
(B) difference in length of the upper limbs
(C) dehydrated intervertebral discs
(D) ipsilateral weakness in gluteal muscles
(E) sciatic nerve lesion

36. Which of the following is NOT true in respect to caudal epidural anesthesia?

(A) A local anesthetic is injected into the sacral hiatus or the posterior sacral foramina.
(B) The anesthetic acts on S2-4 and the coccygeal nerves.
(C) The height to which the anesthetic travels is primarily limited by the amount of fat in the epidural space.
(D) Sensation is lost inferior to the epidural block.
(E) The sacral hiatus is located between the sacral cornua and inferior to the 4th sacral spinous process or median sacral crest.

37. Which of the following is NOT a vertebral problem?

(A) sacralization of L5
(B) lumbarization of S1
(C) lumbarization of T12
(D) lumbar spinal stenosis
(E) hemisacralization of L5

38. Which of the following statements is true?

(A) In spina bifida cystica, the laminae of L5 and possibly S1 do not fuse properly.
(B) In spina bifida occulta, one or more vertebral arches do not develop, allowing meninges and even the spinal cord to herniate.
(C) Paralysis of the limbs and problems with bladder/bowel control may be associated with meningomyelocele.
(D) Some cases of spina bifida result from an improper closure of the neural tube during the 8th week of embryonic development.
(E) A meningocele is a spina bifida associated with brain herniation.

39. Which of the following are derivatives of the epimere?

(A) erector spinae muscles
(B) prevertebral muscles
(C) quadratus lumborum
(D) striated muscles of the anus
(E) sternalis

40. Shortly after week four of development, dorsal primary rami begin to innervate which of the following?

(A) ventral axial skeletal musculature
(B) vertebral joints
(C) skin of the upper limb
(D) sweat glands of the lateral back region
(E) erector pili muscles

DIRECTIONS (Questions 41 through 46): Identify the anatomical features indicated on the art below.

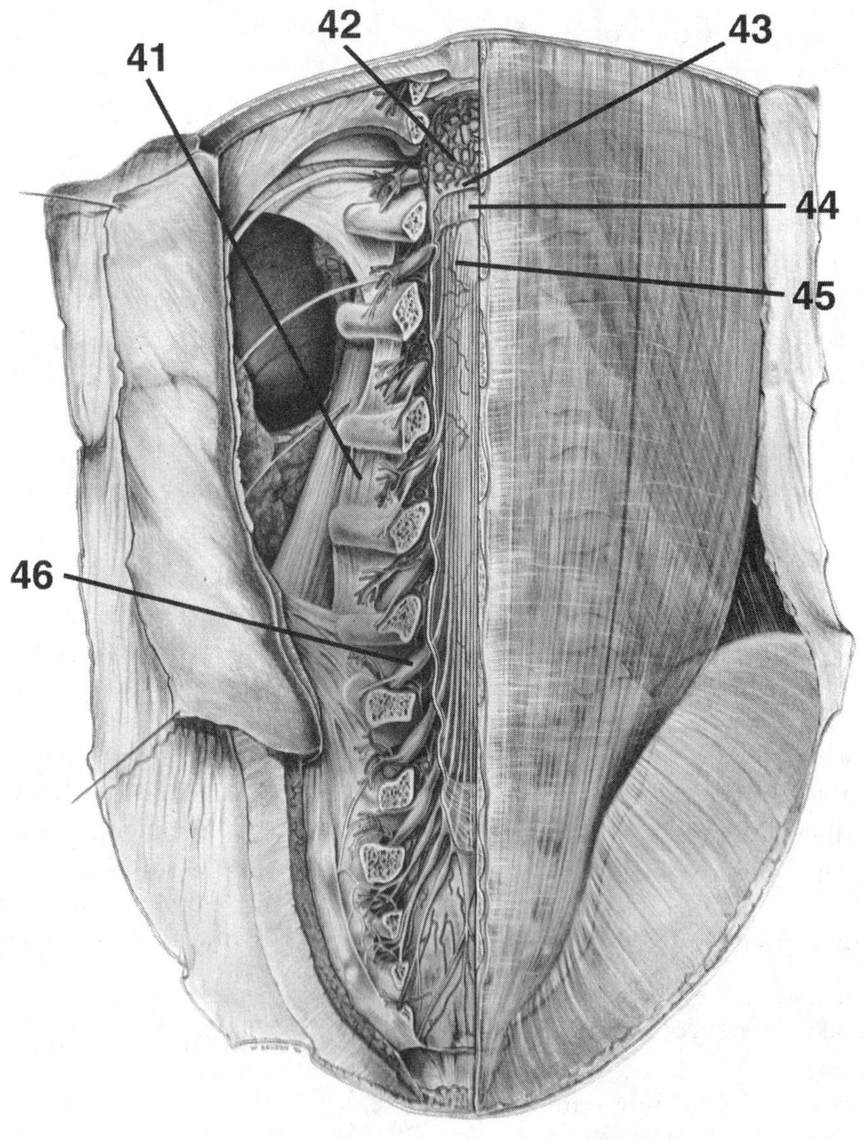

Answers and Explanations

1. **(E)** The vertebral column forms the main part of the axial skeleton *(Moore, p 432).*

2. **(D)** The articular processes contain superior facets directed superoposteriorly *(Moore, p 439).*

3. **(B)** The vertebral foramen is circular and smaller than the foramina of cervical and lumbar vertebrae *(Moore, p 441).*

4. **(A)** The body is massive and kidney-shaped when viewed superiorly. The vertebral foramen is triangular and larger than the foramina in thoracic vertebrae and smaller than those in cervical vertebrae. The transverse processes are long and slender and contain accessory processes. The articular processes contain mammillary processes. The spinous processes are short, thick, and broad *(Moore, p 442).*

5. **(E)** There is no intervertebral disc between the C1 and C2 vertebrae *(Moore, p 451).*

6. **(D)** The posterior longitudinal ligament helps prevent hyperflexion of the vertebral column *(Moore, p 451).*

7. **(A)** Spinal arteries supplying the vertebrae are branches of the vertebral and ascending cervical arteries in the neck, posterior intercostal arteries in the thorax, subcostal and lumbar arteries in the lumbar region, and iliolumbar and lateral and medial sacral arteries in the pelvis *(Moore, p 467).*

8. **(E)** The longissimus, a deep or intrinsic back muscle, is innervated by dorsal rami. All superficial or extrinsic back muscles except the trapezius, which is innervated by the accessory nerve, are innervated by ventral rami *(Moore, p 467).*

9. **(A)** The serratus posterior inferior is an intermediate extrinsic back muscle *(Moore, p 467).*

10. **(A)** The splenius capitis and cervicis, when acting together, extend the head and neck *(Moore, p 468).*

11. **(C)** The erector spinae muscles extend the vertebral column and head when acting bilaterally, control flexion of the back by gradually lengthening their fibers, and bend the vertebral column laterally when acting unilaterally *(Moore, p 470).*

12. **(C)** The spinalis is a member of the erector spinae muscles, which comprise the intermediate muscle layer of the back *(Moore, p 470).*

13. **(C)** The transversospinalis muscles do not elevate the ribs to assist inspiration *(Moore, p 470).*

14. **(E)** The spinalis inserts on spinous processes—not transverse processes. The semispinalis, multifidus, rotatores, and intertransversarii all attach to transverse processes of vertebrae *(Moore, p 470).*

15. **(A)** The longus colli flexes the cervical intervertebral joints but does not bend them laterally *(Moore, p 473).*

16. **(C)** Because of their small size and lack of mechanical advantage, it has been proposed that the rotatores are likely to be organs of proprioception instead of producers of motion *(Moore, p 474).*

17. **(A)** The rectus capitis posterior major arises from the spinous processes of C2 and inserts into the lateral part of the inferior nuchal line. The rectus capitis posterior minor arises from the posterior tubercle of the posterior arch of C1 and inserts into the medial part of the inferior nuchal line. The obliquus capitis inferior arises from the spinous processes of C2 and inserts into the transverse process of C1. The obliquus capitis superior arises from the transverse process of C1 and inserts into the occipital bone between the nuchal lines. The suboccipital muscles are innervated by the dorsal rami of C1 *(Moore, pp 475–476).*

18. **(C)** The longus capitis flexes but does not extend the atlanto-occipital joint *(Moore, p 476).*

19. **(A)** The suboccipital triangle is the deep triangular area between the rectus capitis posterior major and the superior and inferior oblique muscles. The boundaries and course of the suboccipital triangle include the rectus capitis posterior major, the superior oblique, and the inferior oblique. The floor is formed by the atlanto-occipital membrane and posterior arch of C1. The roof is formed by the semispinalis capitis. The suboccipital triangle contains the vertebral artery and suboccipital nerve *(Moore, pp 476–477).*

20. **(D)** The lesser occipital nerve, which is composed of ventral rami of C2 and C3, innervates the skin of the neck and scalp *(Moore, p 477).*

21. **(B)** The spinal cord is enlarged in the lumbosacral region for innervation of the lower limbs *(Moore, p 477).*

22. **(E)** The spinal cord is enlarged in two regions for innervation of the limbs. The tapering end of the spinal cord may terminate as high as T12 or as low as L3. The first cervical nerves lack dorsal roots in 50% of people. The coccygeal nerve may be absent. The terminal filum is the vestigial remnant of the caudal part of the spinal cord that was in the tail of the embryo *(Moore, pp 477–479).*

23. Fat (loose connective tissue) is contained in the extradural (epidural) space *(Moore, p 480).*

24. **(E)** CSF, arachnoid trabeculae, segmental medullary arteries, radicular arteries, and spinal arteries are located in the subarachnoid (leptomeningeal) space. The internal vertebral venous plexus is located in the extradural (epidural) space *(Moore, p 480).*

25. **(E)** The vertebral, ascending cervical, deep cervical, intercostal, lumbar, and lateral sacral arteries give rise to arteries supplying the spinal cord *(Moore, p 486).*

26. **(B)** There are paired posterior spinal arteries *(Moore, p 486).*

27. **(B)** Veins of the spinal cord are distributed in a similar fashion to that of spinal arteries *(Moore, p 486).*

28. **(C)** Typical spinal nerves do not contain parasympathetic fibers *(Moore, p 44–45).*

29. **(C)** The postsynaptic neurons of the parasympathetic nervous system emit acetylcholine *(Moore, p 45).*

30. **(A)** Postsynaptic sympathetic fibers that ultimately innervate the body wall and limbs pass from the sympathetic trunks to adjacent ventral rami through gray rami communicantes *(Moore, p 47).*

31. **(A)** Postsynaptic sympathetic fibers dilate but do not constrict the pupil of the eye *(Moore, p 47).*

32. **(C)** The number of cervical vertebrae is constant at seven *(Moore, p 434).*

33. **(A)** Kyphosis (humpback or hunchback) may result from developmental anomalies as well as from osteoporosis. It is characterized by an abnormal increase in the thoracic curvature with the vertebrae curving posteriorly, resulting in an increase in the anteroposterior diameter of the thorax. Women may develop a temporary lordosis—not kyphosis—during pregnancy *(Moore, p 434).*

34. **(B)** Lordosis is characterized by an abnormal rotation of the pelvis *(Moore, p 434).*

35. **(A)** Scoliosis may be caused by asymmetrical weakness of intrinsic back muscles (myopathic scoliosis), difference in length of the lower limbs, failure of one half of a vertebra to develop, or occasionally habitual standing or sitting in an improper position (habit scoliosis) *(Moore, p 435).*

36. **(C)** The height to which the anesthetic travels is primarily limited by the amount injected and by the position of the patient *(Moore, p 435).*

37. **(C)** Part or all of L5 may fuse with the sacrum (hemisacralization or sacralization). In addition, S1 may separate from the sacrum and fuse with L5. Lumbar stenosis occurs when an intervertebral disc bulges and narrows the vertebral canal in the lumbar region, compressing the spinal nerve roots. T12 is not known to fuse with L1 *(Moore, pp 446–447).*

38. **(C)** Paralysis of the limbs and problems with bladder/bowel control may be present in severe cases of meningomyelocele, which is associated with spina bifida cystica, a condition in which one or more vertebral arches do not develop. In spina bifida occulta, the laminae of L5 and possibly S1 do not fuse properly. Some cases of spina bifida result from an improper closure of the neural tube during the 4th week of embryonic development *(Moore, pp 448–449).*

39. **(A)** Myoblasts of the epimeres form the extensor muscles of the vertebral column *(Sadler, p 190).*

40. **(A)** Dorsal primary rami innervate dorsal axial musculature, vertebral joints, and the skin of the back *(Sadler, p 421).*

41. intertransverse ligament

42. internal vertebral venous plexus

43. dura mater

44. arachnoid layer

45. conus medullaris

46. dorsal root ganglion

CHAPTER 2

The Upper Limb

Questions

DIRECTIONS (Questions 1 through 83): Each of the numbered items or incomplete statements in this section is followed by answers or by completions of the statement. Select the ONE lettered answer or completion that is BEST in each case.

1. Which of the following is NOT true regarding the clavicle?

(A) Its medial end is enlarged where it attaches to the sternum.
(B) Its lateral end is flat where it articulates with the humerus.
(C) The medial two-thirds of the shaft are convex anteriorly.
(D) The clavicle transmits shock from the upper limb to the axial skeleton.
(E) The clavicle is a "long bone" that has no medullary cavity.

2. The trapezius attaches to which of the following regions of the clavicle?

(A) lateral one-third of the clavicle
(B) conoid tubercle
(C) subclavian groove
(D) trapezoid line
(E) quadrangular tubercle

3. Which of the following is true in respect to the scapula?

(A) The spine of the scapula continues laterally as the coracoid process.
(B) The lateral surface of the scapula forms the glenoid cavity.
(C) The acromion is superior to the glenoid cavity and projects anterolaterally.
(D) The scapula is fastened securely to the thoracic cage at the scapulothoracic joint.
(E) The acromioclavicular joint represents the true shoulder joint.

4. Which of the following is NOT included in the condyle of the humerus?

(A) radial, coronoid, and olecranon fossae
(B) epicondyles
(C) trochlea
(D) capitulum
(E) greater tubercle

5. Which of the following is NOT true in respect to the ulna and radius?

(A) The brachialis attaches to the tuberosity of the ulna.
(B) The ulnar styloid process is much larger than the radial styloid process and extends farther distally.
(C) The head of the ulna lies distally, whereas the head of the radius articulates with the humerus.
(D) The ulna is medial to the radius in the anatomical position.
(E) The bodies of these bones are firmly bound together by the interosseous membrane.

6. Which of the following is true regarding the carpus?
   (A) The scaphoid articulates proximally with the ulna and has a tubercle.
   (B) The lunate articulates with the ulna and is broader anteriorly than posteriorly.
   (C) The triquetrum articulates proximally with the articular disc of the distal radioulnar joint.
   (D) The pisiform lies on the palmar surface of the trapezium.
   (E) It is composed of seven bones.

7. Which of the following describes the correct order of the distal row of carpals from lateral to medial?
   (A) triquetrum, trapezoid, capitate, hamate
   (B) trapezoid, trapezium, capitate, hamate
   (C) trapezium, trapezoid, capitate, hamate
   (D) trapezium, triquetrum, capitate, hamate
   (E) scaphoid, lunate, triquetrum, pisiform

8. Which of the following is actually a lateral cutaneous branch of an intercostal nerve, innervating the skin of the medial surface of the arm?
   (A) intercostobrachial nerve
   (B) superior lateral cutaneous nerve of the arm
   (C) inferior lateral cutaneous nerve of the arm
   (D) medial cutaneous nerve of the arm
   (E) lateral pectoral nerve

9. Which of the following is NOT a branch of the radial nerve?
   (A) posterior cutaneous nerve of the arm
   (B) posterior cutaneous nerve of the forearm
   (C) inferior lateral cutaneous nerve of the arm
   (D) superior lateral cutaneous nerve of the arm
   (E) posterior interosseous nerve

10. Which of the following is NOT an anterior thoracoappendicular muscle?
   (A) pectoralis major
   (B) pectoralis minor
   (C) deltoid
   (D) subclavius
   (E) serratus anterior

11. Which of the following best describes the action of the pectoralis minor?
   (A) stabilizes scapula by drawing it inferiorly and anteriorly against thoracic wall
   (B) anchors and depresses clavicle
   (C) adducts and medially rotates humerus
   (D) rotates scapula
   (E) flexes humerus

12. Which of the following muscles attaches to the coracoid process of the scapula?
   (A) pectoralis minor
   (B) triceps brachii
   (C) brachialis
   (D) pectoralis major
   (E) subclavius

13. All of the following are medial rotators of the arm EXCEPT
   (A) latissimus dorsi
   (B) teres major
   (C) subscapularis
   (D) infraspinatus
   (E) anterior part of deltoid

14. What muscles are necessary to raise the arm above the shoulder?
   (A) first the supraspinatus, next the deltoid, and then the serratus anterior
   (B) first the deltoid, next the supraspinatus, and then the serratus anterior
   (C) first the supraspinatus, next the serratus anterior, and then the deltoid
   (D) first the serratus anterior, next the deltoid, and then the supraspinatus
   (E) first the deltoid, next the serratus anterior, and then supraspinatus

15. Which of the following is innervated by the dorsal scapular nerve?
   (A) serratus anterior
   (B) rhomboid major and minor

(C) erector spinae
(D) subscapularis
(E) supraspinatus

16. Which of the following is an extrinsic shoulder muscle?

(A) deltoid
(B) teres major
(C) levator scapulae
(D) teres minor
(E) supraspinatus

17. Which of the following is true in respect to the trapezius?

(A) It is innervated by the dorsal scapular nerve.
(B) Its superior fibers retract the scapula.
(C) Its middle fibers elevate the scapula.
(D) Its inferior fibers retract the scapula.
(E) Its superior and inferior fibers act together in rotating the scapula on the thoracic wall.

18. A patient is asked to place the hands posteriorly on the hips and to push the elbows posteriorly against resistance. Which muscle is being tested?

(A) levator scapulae
(B) rhomboid
(C) trapezius
(D) latissimus dorsi
(E) serratus anterior

19. Which rotator cuff muscle does NOT rotate the humerus?

(A) supraspinatus
(B) infraspinatus
(C) teres minor
(D) subscapularis
(E) teres major

20. The axillary nerve innervates which of the following muscles?

(A) coracobrachialis
(B) teres minor
(C) teres major
(D) subscapularis
(E) levator scapulae

21. Which of the following is NOT contained in the axilla?

(A) axillary blood vessels
(B) lymph nodes
(C) trunks and divisions of the brachial plexus
(D) axillary nerve
(E) lymph nodes

22. Which of the following is most correct?

(A) The subscapular artery arises from the third part of the axillary artery and contributes to blood supply of muscles near the scapula and humerus.
(B) The second part of the axillary artery typically contains two branches—the thoracoacromial artery and the superior thoracic artery.
(C) The first part of the axillary artery lies posterior to the pectoralis minor.
(D) The thoracoacromial artery supplies the pectoral muscles, axillary lymph nodes, and most importantly the lateral part of the mammary gland in women.
(E) The lateral thoracic artery divides into four branches, the acromial, deltoid, pectoral, and clavicular.

23. Which of the following is NOT correct?

(A) The brachial plexus is formed by the union of the ventral rami of C5 through T1.
(B) The roots of the brachial plexus and the subclavian artery pass through the gap between the anterior and middle scalene muscles.
(C) Gray rami contribute sympathetic fibers to each root.
(D) Each of the three trunks of the brachial plexus divide into anterior and posterior divisions.
(E) The cords of the brachial plexus surround the brachial artery.

24. Which of the following is NOT a supraclavicular branch of the brachial plexus?

(A) dorsal scapular nerve
(B) lateral pectoral nerve
(C) long thoracic nerve
(D) nerve to the subclavius
(E) suprascapular nerve

25. Which of the following is true regarding the quadrangular space?

(A) It is bounded superiorly by the teres major.
(B) It is bounded inferiorly by the subscapularis and teres minor.
(C) It is bounded medially by the humerus and laterally by the long head of the triceps.
(D) It contains the posterior circumflex humeral artery and the axillary nerve.
(E) Brachial plexus herniations occur here.

26. Which of the following is NOT innervated by the suprascapular nerve?

(A) supraspinatus
(B) infraspinatus
(C) glenohumeral joint
(D) skin over superior part of scapula
(E) shoulder joint

27. Which of the following is NOT a branch of the posterior cord of the brachial plexus?

(A) upper and lower subscapular nerves
(B) thoracodorsal nerve
(C) axillary nerve
(D) radial nerve
(E) long thoracic nerve

28. Which of the following is NOT true in respect to the brachialis?

(A) Its origin is the distal half of the anterior surface of the humerus.
(B) Its insertion is the coronoid process and tuberosity of the ulna.
(C) It flexes the forearm in all positions.
(D) It is primarily innervated by the musculocutaneous nerve, but some of its lateral part is innervated by a branch of the radial nerve.
(E) It crosses two joints.

29. A patient is asked to abduct the arm 90 degrees and then to extend the flexed forearm against resistance. Which muscle is being tested?

(A) triceps brachii
(B) brachialis
(C) coracobrachialis
(D) biceps brachii
(E) supinator

30. The deep artery of the arm accompanies which of the following before passing around the body of the humerus?

(A) radial nerve
(B) musculocutaneous nerve
(C) median nerve
(D) ulnar nerve
(E) axillary nerve

31. Which muscle assists in extension of the forearm, resists abduction of the ulna during pronation of the forearm, and tenses the capsule of the elbow joint so that it is not pinched when the joint is extended?

(A) anconeus
(B) triceps brachii
(C) coracobrachialis
(D) brachialis
(E) biceps brachii

32. Which of the following is a branch of the brachial artery?

(A) anterior and posterior circumflex humeral arteries
(B) deltoid artery
(C) superior and inferior ulnar collateral arteries
(D) thoracoacromial artery
(E) anterior and posterior ulnar recurrent arteries

33. Which of the following nerves supply NO branches to the arm?

(A) musculocutaneous and median
(B) radial and ulnar
(C) median and ulnar
(D) median and radial
(E) musculocutaneous and radial

34. Which of the following nerves is correctly paired with its cutaneous branch?

(A) median nerve and medial antebrachial cutaneous nerve
(B) musculocutaneous nerve and lateral antebrachial cutaneous nerve
(C) ulnar nerve and posterior antebrachial cutaneous nerve
(D) median nerve and medial brachial cutaneous nerve
(E) radial nerve and superior lateral brachial cutaneous nerve

35. The cubital fossa does NOT contain which of the following?

(A) terminal part of the brachial artery
(B) deep accompanying veins of the arteries
(C) median nerve
(D) biceps brachii tendon
(E) ulnar nerve

36. A patient is unable to flex the arm and forearm. Where is the lesion likely to be?

(A) ventral rami of C3–C4
(B) ventral rami of C5–C6–C7
(C) dorsal rami of C6–C7–C8
(D) ventral rami of C8–T1
(E) dorsal rami of T1

37. The radial nerve innervates muscles in the extensor compartment of the forearm, but it also innervates the following flexor:

(A) brachioradialis
(B) pronator teres
(C) palmaris longus
(D) pronator quadratus
(E) palmaris longus

38. Which muscle does NOT cross the elbow joint?

(A) flexor pollicis longus
(B) pronator teres
(C) flexor carpi radialis
(D) flexor carpi ulnaris
(E) flexor digitorum superficialis

39. The ulnar nerve innervates which of the following muscles in the flexor compartment?

(A) the medial part of the flexor digitorum superficialis
(B) flexor carpi radialis
(C) pronator quadratus
(D) pronator teres
(E) the medial part of flexor digitorum profundus

40. The radial artery lies just lateral to the tendon of which muscle?

(A) pronator teres
(B) flexor carpi radialis
(C) palmaris longus
(D) flexor carpi ulnaris
(E) flexor digitorum superficialis

41. The palmaris longus tendon is a useful guide to which nerve at the wrist?

(A) anterior interosseous nerve
(B) posterior interosseous nerve
(C) median nerve
(D) ulnar nerve
(E) radial nerve

42. To pronate the forearm, which of the following must occur?

(A) The pronator quadratus initiates pronation, assisted later by the pronator teres.
(B) The pronator teres initiates pronation, assisted later by the pronator quadratus.
(C) The anconeus initiates pronation, assisted later by the pronator teres.
(D) The pronator quadratus initiates pronation, assisted later by the anconeus.
(E) The ulnar nerve must be used.

43. The extensor carpi radialis longus tendon is crossed by which two muscles?

(A) abductor pollicis longus and extensor pollicis longus
(B) extensor indicis and extensor digitorum
(C) extensor digitorum and extensor pollicis brevis
(D) abductor pollicis longus and extensor pollicis brevis
(E) extensor indicis and extensor carpi radialis brevis

44. Which of the following is true in respect to the supinator?

(A) It is innervated by the ulnar nerve.
(B) It supinates the forearm by rotating the ulna.
(C) It forms the floor of the cubital fossa along with the brachioradialis.
(D) It supinates the forearm when the forearm is already flexed.
(E) It rotates the radius to turn the palm anteriorly.

45. Which of the following does NOT take an origin from the lateral epicondyle of the humerus?

(A) extensor carpi radialis brevis
(B) extensor carpi ulnaris
(C) abductor pollicis longus
(D) supinator
(E) extensor digiti minimi

46. Which of the following is correctly paired with its nerve?

(A) flexor pollicis longus and anterior interosseous nerve
(B) flexor digitorum profundus and anterior interosseous nerve
(C) extensor carpi radialis longus and posterior interosseous nerve
(D) brachioradialis and posterior interosseous nerve
(E) abductor pollicis longus and anterior interosseous nerve

47. Which of the following is true in respect to the anatomical snuff box?

(A) It is bounded anteriorly by the tendons of the extensor pollicis longus.
(B) It is bounded posteriorly by the tendons of the abductor pollicis longus and extensor pollicis brevis.
(C) The radial artery lies in the floor of the snuff box.
(D) The scaphoid and triquetrum can be palpated within the snuff box.
(E) The snuff box is visible when the thumb is fully flexed.

48. Which of the following does NOT abduct the hand at the wrist joint?

(A) flexor carpi radialis
(B) extensor carpi radialis longus
(C) extensor carpi radialis brevis
(D) abductor pollicis longus
(E) palmaris longus

49. Which of the following is derived from the radial artery?

(A) dorsal and palmar carpal arteries
(B) common interosseous artery
(C) anterior interosseous artery
(D) poster interosseous artery
(E) ulnar recurrent artery

50. The median nerve does which of the following?

(A) innervates the elbow joint with articular branches
(B) innervates the medial half of the flexor digitorum profundus
(C) innervates the hypothenar muscles
(D) innervates lumbricals 3 and 4
(E) innervates the skin of the dorsum of the hand

51. The ulnar nerve does NOT do which of the following?

(A) innervate the elbow joint with articular branches
(B) innervate the flexor carpi ulnaris

(C) innervate the skin on the lateral part of the palm and dorsum of the hand
(D) innervate the adductor pollicis
(E) innervate the dorsal and palmar interossei

52. The radial nerve does NOT do which of the following?

(A) give a superficial branch that innervates the dorsum of the hand
(B) innervate the brachioradialis and extensor carpi radialis longus
(C) give a deep branch that innervates the extensor carpi radialis brevis and supinator
(D) give a posterior interosseous branch that innervates all remaining extensor muscles in the posterior compartment of the forearm
(E) innervate the glenohumeral joint

53. Which of the following is NOT true in respect to the flexor pollicis brevis?

(A) It is located medial to the abductor pollicis brevis.
(B) It flexes the thumb at the carpometacarpal joint.
(C) It flexes the thumb at the metacarpophalangeal joint.
(D) Its tendon typically contains a sesamoid bone.
(E) It is innervated by C5–C6.

54. Which of the following is true in respect to the palmaris brevis?

(A) It aids the palmaris longus in tightening the palmar aponeurosis.
(B) It is innervated by the median nerve.
(C) It is in the hypothenar compartment.
(D) It covers and protects the radial artery.
(E) It wrinkles the skin of the hypothenar eminence and deepens the hollow of the palm.

55. The recurrent branch of the median nerve does NOT innervate which of the following?

(A) abductor pollicis brevis
(B) adductor pollicis
(C) flexor pollicis brevis
(D) opponens pollicis
(E) The recurrent branch of the median nerve innervates all of the above.

56. Which of the following muscles is correctly matched with the accompanying description?

(A) lumbricals 1 and 2 . . . bipennate
(B) lumbricals 3 and 4 . . . unipennate
(C) dorsal interossei 1–4 . . . bipennate
(D) palmar interossei 1–3 . . . bipennate
(E) deltoid . . . bipennate

57. The deep branch of the ulnar does NOT innervate which of the following?

(A) abductor digiti minimi
(B) flexor digiti minimi brevis
(C) lumbricals 1 and 2
(D) dorsal interossei 3 and 4
(E) palmar interossei 1 and 2

58. The carpal tunnel does NOT contain which of the following?

(A) median nerve
(B) four tendons of the flexor digitorum superficialis
(C) four tendons of the flexor digitorum profundus
(D) the tendon of the flexor pollicis longus
(E) ulnar nerve

59. The sternoclavicular joint . . .

(A) . . . is a saddle type synovial joint but functions as a ball-and-socket joint.
(B) . . . is supplied by lateral thoracic and thoracoacromial arteries.
(C) . . . is innervated by the lateral and medial pectoral nerves.
(D) . . . is the articulation of the clavicle and gladiolus of the sternum.
(E) . . . dislocates easily.

60. Which of the following is true in respect to the acromioclavicular joint?

(A) It is a saddle-type synovial joint.
(B) It is strengthened by the coracohumeral and transverse humeral ligaments.
(C) It is supplied by the lateral thoracic arteries.
(D) It is innervated by the nerve to the subclavius.
(E) When dislocated, it is often referred to as a "separated shoulder."

61. Which of the following flexes the arm at the glenohumeral joint?

(A) deltoid (posterior part)
(B) pectoralis major
(C) latissimus dorsi
(D) subscapularis
(E) infraspinatus

62. In respect to movement of the arm at the glenohumeral joint, which of the following movements is correctly paired with its prime mover?

(A) extension . . . deltoid (posterior part)
(B) abduction . . . pectoralis major and latissimus dorsi
(C) adduction . . . deltoid
(D) medial rotation . . . infraspinatus
(E) lateral rotation . . . subscapularis

63. Which of the following is true in respect to the elbow joint?

(A) It is a plane type of synovial joint.
(B) It is strengthened by the radial and ulnar cruciate ligaments.
(C) It is supplied by the cephalic and basilic arteries.
(D) It is innervated by the median and axillary nerves.
(E) It is surrounded by the intratendinous olecranon bursa, the subtendinous olecranon bursa, and the subcutaneous olecranon bursa.

64. Which of the following joints is paired correctly with its type?

(A) proximal and distal radioulnar joints . . . condyloid type of synovial joint
(B) radiocarpal joint . . . pivot type of synovial joint
(C) intercarpal joints . . . plane type of synovial joints
(D) metacarpophalangeal joints . . . hinge type of synovial joints
(E) interphalangeal joints . . . condyloid type of synovial joints

65. All carpometacarpal and intermetacarpal joints are plane types of synovial joints EXCEPT for

(A) the carpometacarpal joint of the thumb.
(B) the carpometacarpal joint of the fifth metacarpal.
(C) the carpometacarpal joint of the third metacarpal.
(D) the intermetacarpal joint of the 4th and 5th metacarpals.
(E) the intermetacarpal joint of the 1st and 2nd metacarpals.

66. Which of the following is NOT true in respect to the clavicle?

(A) The clavicle varies more in shape than most other long bones.
(B) The clavicle can be pierced by a branch of the supraclavicular nerve.
(C) The clavicle is thicker and more curved in manual workers.
(D) The right clavicle is stronger than the left and is usually shorter.
(E) The clavicle is a compact bone.

67. Fractures of the scapula typically involve

(A) the acromion.
(B) the coracoid process.
(C) the spine.
(D) the inferior angle.
(E) the suprascapular notch.

68. Which of the following parts of the humerus is matched correctly with the nerve with which it is in direct contact?

(A) distal end of humerus . . . radial nerve
(B) surgical neck . . . musculocutaneous nerve
(C) radial groove . . . musculocutaneous nerve
(D) medial epicondyle . . . ulnar nerve
(E) scapular notch . . . suprascapular nerve

69. "Winging" of the scapula is most likely caused by which of the following?

(A) a lesion to the long thoracic nerve
(B) a lesion to the thoracodorsal nerve
(C) injury to the suprascapular nerve
(D) damage to the dorsal scapular nerve
(E) damage to the upper and lower subscapular nerves

70. Which of the following is correct regarding the triangle of auscultation?

(A) Its borders are the latissimus dorsi, scapula, and trapezius.
(B) It is a good location to hear heart murmurs.
(C) The 8th and 9th ribs and the 8th intercostal space are subcutaneous here.
(D) It is a location of back trauma.
(E) It is a location for dorsal rami to pass to the superficial back.

71. A patient cannot raise the trunk (as in climbing). What is most likely the problem?

(A) damage to the ventral rami of C5–C6–C7
(B) paralysis of the latissimus dorsi
(C) injury to the dorsal scapular nerve
(D) damage to the dorsal rami of C8–T1
(E) injury to the axillary nerve

72. The scapula on one side of a patient is located farther from the midline than that on the normal side. What might be the problem?

(A) paralysis of the rhomboids on one side
(B) injury to the long thoracic nerve
(C) a lesion of C7–C8
(D) dislocated shoulder
(E) separated shoulder

73. The axillary nerve is damaged. What is the likely result?

(A) The teres major atrophies.
(B) The rounded contour of the shoulder disappears.
(C) A loss of sensation may occur in the lateral forearm.
(D) The patient may lose the ability to adduct the arm.
(E) The patient may exhibit "wrist-drop."

74. Which of the following is true regarding rotator cuff injuries?

(A) Injury or disease may damage the rotator cuff, causing instability of the acromioclavicular joint.
(B) The supraspinatus tendon is the most commonly torn part of the rotator cuff.
(C) The teres major takes the longest to rehabilitate of the rotator cuff muscles.
(D) The injuries occur when the muscles pull away from their origin on the acromion.
(E) Acute tears are common in young persons.

75. A patient has been thrown from a motorcycle, landing on the shoulder such that the neck and shoulder are widely separated. You suspect an upper brachial plexus injury. What signs do you expect?

(A) "clawhand"
(B) paralysis of flexor carpi ulnaris, flexor digitorum superficialis, and flexor digitorum profundus
(C) adducted shoulder, medially rotated arm, and extended elbow
(D) loss of sensation in the medial forearm
(E) "wrist-drop"

76. A patient exhibits "clawhand." What might have happened?

(A) upper brachial plexus injury
(B) acute brachial plexus neuritis
(C) compression of the cords of the brachial plexus
(D) lower brachial plexus injury
(E) damage to dorsal rami that send fibers to the brachial plexus

77. A patient receives a knife wound to the axilla. What problems do you expect?

(A) damage to the axillary nerve
(B) paralysis of the coracobrachialis, biceps, and brachialis
(C) inability to extend the wrist and digits at the metacarpophalangeal joints
(D) loss of sensation on the medial surface of the arm
(E) "clawhand"

78. A patient tries to make a fist, but digits 2 and 3 remain partially extended. What nerve is injured?

(A) ulnar nerve
(B) radial nerve
(C) median nerve
(D) musculocutaneous nerve
(E) axillary nerve

79. Which of the following is true in respect to ulnar nerve injuries?

(A) The injury often occurs where the nerve passes posterior to the medial epicondyle of the humerus.
(B) The patient experiences numbness and tingling on the lateral part of the palm and the thumb.
(C) The patient may exhibit "waiter's tip hand."
(D) Patients have difficulty because they cannot flex their first, second, and third digits at the DIP joints.
(E) Power of abduction is impaired, and when the patient attempts to flex the wrist, the flexor carpi ulnaris brings the hand to the medial side.

80. Which limb defect is correctly matched with its definition?

(A) meromelia . . . complete absence of one or more extremities
(B) phocomelia . . . all segments of extremities are present but abnormally short
(C) micromelia . . . partial absence of one or more extremities
(D) amelia . . . long bones are absent, and small hands or feet are attached to the trunk by short, irregular bones
(E) cleft hand (lobster claw deformity) . . . absent third metacarpal, fusion of digits 1–2 and 4–5

81. Syndactylyl involves

(A) extra fingers or toes.
(B) absence of a digit or limb.
(C) abnormal fusion of fingers and toes.
(D) small hands or feet being attached to trunk by short bones instead of long bones.
(E) congenital dislocation of glenohumeral joint.

82. Which of the following is NOT correct?

(A) During development, dorsal cells organize as the epimere and ventral cells organize as the hypomere.
(B) Dorsal rami innervate muscles derived from the epimere.
(C) Ventral rami innervate muscles derived from the hypomere.
(D) Myoblasts of the hypomere form the extensor muscles of the vertebral column.
(E) Somites and somitomeres form the musculature of the limbs.

83. A patient in surgery has no pectoralis major. What do you suspect?

(A) trauma
(B) dominant pectoralis minor
(C) drug-induced muscle hypoplasia
(D) atrophy of the muscle
(E) congenital absence of the muscle

DIRECTIONS (Questions 84 through 88): Identify the anatomical features indicated on the art below.

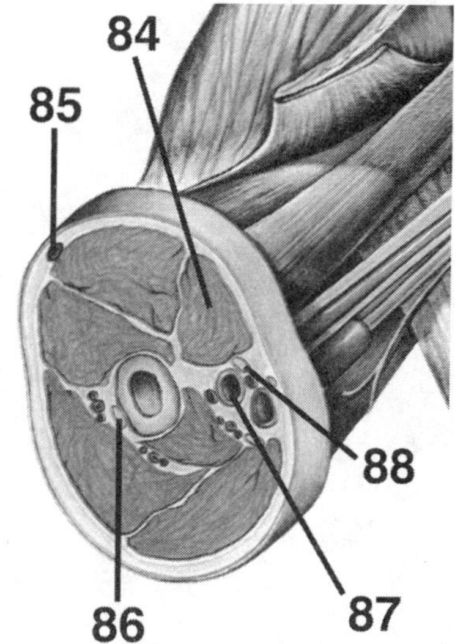

DIRECTIONS (Questions 89 through 93): Identify the anatomical features indicated on the art below.

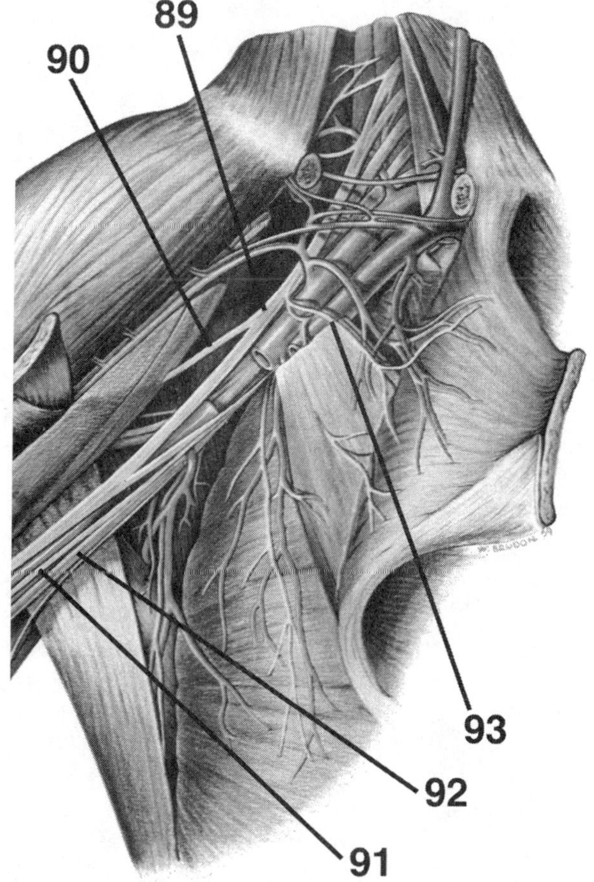

DIRECTIONS (Questions 94 through 98): Identify the anatomical features indicated on the art below.

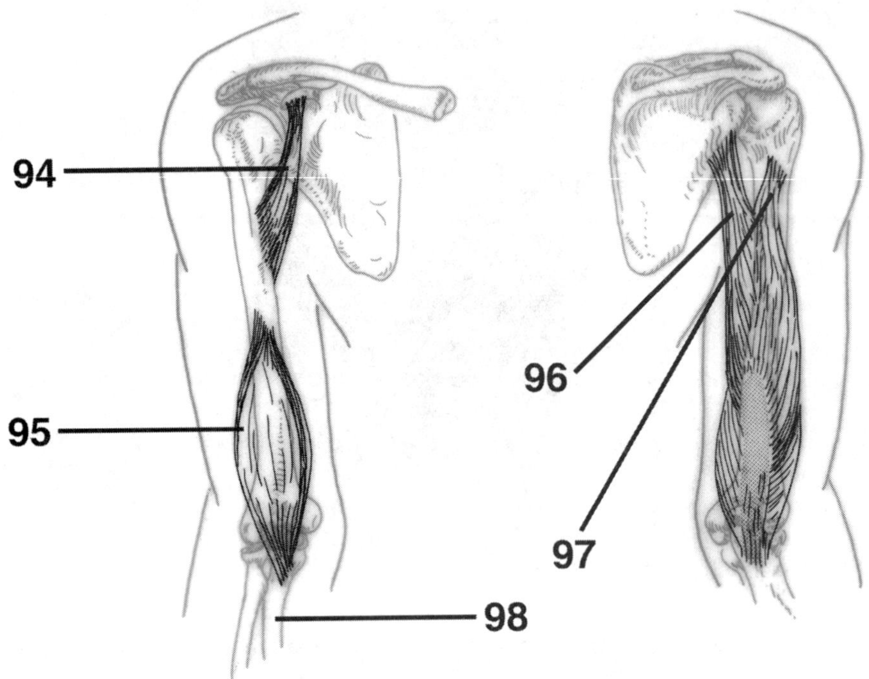

Answers and Explanations

1. **(B)** The lateral end of the clavicle is flat where it articulates with the acromion at the acromioclavicular (AC) joint *(Moore, pp 665–666).*

2. **(A)** The trapezius attaches to the lateral third of the clavicle, acromion, and spine of the scapula. The deltoid attaches to the deltoid tubercle, the conoid ligament attaches to the conoid tubercle, the subclavius attaches to the subclavian groove, and the trapezoid ligament attaches to the trapezoid line *(Moore, pp 666, 691).*

3. **(B)** The lateral surface of the scapula forms the glenoid cavity, superior to which the coracoid process projects anterolaterally. The glenohumeral joint itself represents the true shoulder joint, whereas the scapulothoracic joint, which is a conceptual joint, is a location where the scapula moves easily on the thoracic wall. The spine of the scapula continues laterally as the acromion *(Moore, pp 668–669).*

4. **(E)** The condyle of the humerus (the distal end) includes the epicondyles, trochlea, capitulum, and the three fossae (radial, coronoid, and radial) *(Moore, p 670).*

5. **(B)** The radial styloid process is much larger than the ulnar styloid process and extends farther distally *(Moore, p 671–672).*

6. **(C)** The scaphoid articulates proximally with the radius and has a large tubercle. The lunate articulates with the radius and is broader anteriorly than posteriorly. The triquetrum articulates proximally with the articular disc of the distal radioulnar joint. The pisiform lies on the palmar surface of the triquetrum *(Moore, p 674).*

7. **(C)** From lateral to medial, the four bones of the distal row of carpals are the trapezium, trapezoid, capitate, and hamate *(Moore, p 674).*

8. **(A)** The intercostobrachial nerve is the lateral cutaneous branch of the second intercostal nerve from T2, innervating the skin of the medial surface of the arm *(Moore, p 684).*

9. **(D)** The posterior cutaneous nerve of the arm, posterior cutaneous nerve of the forearm, and inferior lateral cutaneous nerve of the arm are branches of the radial nerve. The superior lateral cutaneous nerve is a branch of the axillary nerve *(Moore, p 684).*

10. **(C)** The pectoralis major, pectoralis minor, subclavius, and serratus anterior are anterior thoracoappendicular muscles. The deltoid is a scapulohumeral (shoulder) muscle *(Moore, pp 688, 691).*

11. **(A)** The pectoralis minor stabilizes the scapula by drawing it inferiorly and anteriorly against the thoracic wall *(Moore, p 688).*

12. **(A)** The pectoralis minor, biceps brachii (short head), and coracobrachialis attach to the coracoid process of the scapula *(Moore, p 688).*

13. **(D)** The latissimus dorsi, teres major, and subscapularis medially rotate the arm. The infraspinatus and teres minor rotate the arm laterally. The deltoid is unique in that its anterior part rotates the arm medially, and its posterior part rotates the arm laterally *(Moore, p 691).*

14. **(A)** The supraspinatus initiates abduction of the arm. The deltoid becomes fully effective as

an abductor following the initial 15 degrees of abduction. The serratus anterior rotates the scapula, elevating its glenoid cavity so that the arm can be raised above the shoulder *(Moore, pp 695, 696, 689).*

15. **(B)** The dorsal scapular nerve innervates the levator scapulae, rhomboid major, and rhomboid minor *(Moore, p 691).*

16. **(C)** The trapezius, latissimus dorsi, levator scapulae, and rhomboids are extrinsic shoulder muscles. The deltoid, teres major, supraspinatus, infraspinatus, teres minor, and subscapularis are intrinsic shoulder muscles *(Moore, pp 691–692).*

17. **(E)** The trapezius, innervated by the spinal root of the accessory nerve (XI), is composed of three types of fibers. Its superior fibers elevate the scapula, its middle fibers retract the scapula, and its inferior fibers depress the scapula. Its superior and inferior fibers act together in rotating the scapula on the thoracic wall *(Moore, p 694).*

18. **(B)** To test the rhomboids, the patient is asked to place the hands posteriorly on the hips and to push the elbows posteriorly against resistance *(Moore, p 695).*

19. **(A)** The supraspinatus is the only rotator cuff muscle that does not rotate the humerus *(Moore, pp 697–698).*

20. **(B)** The axillary nerve innervates both the deltoid and the teres minor *(Moore, p 691).*

21. **(C)** The axilla contains axillary blood vessels, lymph nodes, the cords and branches of the brachial plexus, and the axillary nerve. The trunks and divisions are found superior to the axilla in the neck *(Moore, p 699).*

22. **(A)** The subscapular artery arises from the third part of the axillary artery and contributes to blood supply of muscles near the scapula and humerus *(Moore, p 701).*

23. **(E)** The brachial plexus is formed by the union of the ventral rami of C5 through T1. The roots of the brachial plexus and the subclavian artery pass through the gap between the anterior and middle scalene muscles. Gray rami contribute sympathetic fibers to each root. The roots of the brachial plexus form three trunks, each of which divides into anterior and posterior divisions. The cords of the brachial plexus surround the axillary artery *(Moore, p 708).*

24. **(B)** The dorsal scapular nerve, long thoracic nerve, nerve to the subclavius, and suprascapular nerve are supraclavicular branches of the brachial plexus, whereas the lateral pectoral nerve is an infraclavicular branch, originating from the lateral cord *(Moore, pp 708–709).*

25. **(D)** The quadrangular space is bounded superiorly by the subscapularis and teres minor, inferiorly by the teres major, medially by the long head of triceps, and laterally by the humerus. It contains the axillary nerve and the posterior circumflex humeral artery *(Moore, p 711).*

26. **(D)** The suprascapular nerve innervates the supraspinatus, infraspinatus, and glenohumeral (shoulder) joint *(Moore, p 710).*

27. **(E)** The posterior cord gives rise to the upper and lower subscapular nerves, thoracodorsal nerve, axillary nerve, and radial nerve. The long thoracic nerve originates from C5–C6–C7 *(Moore, pp 711).*

28. **(E)** The brachialis originates from the distal half of the anterior surface of the humerus and inserts on the coronoid process and tuberosity of the ulna. It crosses one joint, flexing the forearm in all positions. While it is primarily innervated by the musculocutaneous nerve, some of its lateral part is innervated by a branch of the radial nerve *(Moore, p 722).*

29. **(A)** To test the triceps brachii, the arm is abducted 90 degrees and then the flexed forearm is extended against resistance *(Moore, p 724).*

30. **(A)** The deep artery of the arm accompanies the radial nerve through the radial groove and passes around the body of the humerus *(Moore, p 728).*

31. **(A)** The anconeus assists in extension of the forearm, resists abduction of the ulna during pronation of the forearm, and tenses the capsule of the elbow joint so that it is not pinched when the joint is extended *(Moore, p 724).*

32. **(C)** The axillary artery gives rise to the superior thoracic, thoracoacromial, lateral thoracic, subscapular, and anterior and posterior circumflex humeral arteries. The brachial artery gives rise to the deep artery of the arm, the nutrient humeral artery, and the superior and inferior collateral arteries. The ulnar artery gives rise to the anterior and posterior ulnar recurrent, common interosseous, anterior and posterior interosseous, and dorsal and palmar carpal branch arteries *(Moore, pp 699, 727–728, 750).*

33. **(C)** The median and ulnar nerves supply no branches to the arm *(Moore, p 730).*

34. **(B)** The median nerve gives rise to a palmar cutaneous branch; the ulnar nerve as well has a palmar cutaneous branch. The radial nerve gives rise to the posterior brachial cutaneous nerve, the posterior antebrachial cutaneous nerve, the inferior lateral brachial cutaneous nerve, and a superficial branch that innervates the dorsum of the hand and the digits. The musculocutaneous nerve continues as the lateral antebrachial cutaneous nerve. The axillary nerve gives rise to the superior lateral brachial cutaneous nerve. The medial cord of the brachial plexus gives rise to the medial brachial cutaneous nerve and the medial antebrachial cutaneous nerve. The supraclavicular nerves (from C3–C4) and the intercostobrachial nerve (from T2) also contribute to the cutaneous innervation of the arm *(Moore, pp 684, 758–759).*

35. **(E)** The cubital fossa contains the terminal part of the brachial artery (and the beginning of the ulnar and radial arteries), deep accompanying veins, the median nerve, and the biceps brachii tendon. In the tissue superficial to the fossa are the median cubital vein and medial and lateral antebrachial cutaneous nerve. The deep and superficial branches of the radial nerve are within the floor of the fossa *(Moore, pp 731–732).*

36. **(B)** A patient who is unable to flex the arm and forearm is likely to have a lesion in the ventral rami of C5, C6, and C7. The biceps brachii and brachialis receive fibers from C5 and C6, and the coracobrachialis receives fibers from C5, C6, and C7. C6 is the main source of fibers for each *(Moore, p 722).*

37. **(A)** The brachioradialis is a flexor of the forearm, but it is located in the extensor compartment and is innervated by the radial nerve *(Moore, p 734).*

38. **(A)** The superficial muscles (pronator teres, flexor carpi radialis, palmaris longus, flexor carpi ulnaris, and flexor digitorum superficialis) cross the elbow joint. The deep muscles (flexor digitorum profundus, flexor pollicis longus, and pronator quadratus) do not *(Moore, p 734).*

39. **(E)** All muscles in the anterior compartment of the forearm are innervated by the median nerve, except for the flexor carpi ulnaris and the medial part of the flexor digitorum profundus, which are innervated by the ulnar nerve *(Moore, pp 736–737).*

40. **(B)** The radial artery lies lateral to the tendon of the flexor carpi radialis *(Moore, p 737).*

41. **(C)** The palmaris longus tendon is a guide for locating the median nerve at the wrist *(Moore, p 737).*

42. **(A)** To pronate the forearm, the pronator quadratus initiates pronation, assisted later by the pronator teres *(Moore, p 741).*

43. **(D)** The extensor carpi radialis longus tendon is crossed by the abductor pollicis longus and extensor pollicis brevis *(Moore, p 745).*

44. **(E)** The supinator, which forms the floor of the cubital fossa along with the brachioradialis, is innervated by the deep branch of the radial nerve. It supinates the forearm by rotating the radius. The biceps brachii also supinates the forearm when the forearm is already flexed *(Moore, p 746).*

45. **(C)** The abductor pollicis longus originates from the posterior surfaces of the ulna, radius, and interosseous membrane. The following muscles take at least one of their origins from the lateral epicondyle of the humerus: extensor carpi radialis brevis, extensor digitorum, extensor digiti minimi, extensor carpi ulnaris, and supinator *(Moore, p 742).*

46. **(A)** The flexor pollicis longus is innervated by the anterior interosseous nerve from the median nerve *(Moore, pp 736–737, 742–743).*

47. **(C)** The snuff box is bounded anteriorly by the tendons of the abductor pollicis longus and extensor pollicis brevis. Posteriorly it is bounded by the tendon of the extensor pollicis longus. The radial artery can be felt in the floor, along with the radial styloid process, first metacarpal, scaphoid, and trapezium *(Moore, p 749).*

48. **(E)** The hand is abducted at the wrist joint by the flexor carpi radialis, abductor pollicis longus, extensor carpi radialis longus, and extensor carpi radialis brevis *(Moore, pp 736, 742).*

49. **(A)** The radial artery gives rise to the radial recurrent artery as well as dorsal and palmar carpal branches. The ulnar artery gives rise to the common interosseous artery, anterior and posterior interosseous arteries, anterior and posterior ulnar recurrent arteries, and dorsal and palmar carpal branches *(Moore, p 750).*

50. **(A)** The median nerve assists in the innervation of the elbow joint and gives muscular branches to pronator teres, flexor carpi radialis, palmaris longus, and flexor digitorum superficialis. The median nerve also has an anterior interosseous branch that innervates the lateral part of the flexor digitorum profundus, flexor pollicis longus, and pronator quadratus. The recurrent branch of the median nerve innervates the thenar muscles, and the palmar cutaneous branch innervates the skin of the lateral part of the palm *(Moore, pp 757–759).*

51. **(C)** The ulnar nerve gives rise to articular branches that innervate the elbow joint and muscular branches that innervate the flexor carpi ulnaris and medial half of the flexor digitorum profundus. The palmar cutaneous branch innervates the skin of the medial part of the palm, and the dorsal cutaneous branch innervates the posterior surface of the medial part of the hand and digits. The deep branch innervates the hypothenar muscles, adductor pollicis, interossei, and the 3rd and 4th lumbricals *(Moore, pp 759–760).*

52. **(E)** The radial nerve gives a superficial branch that innervates skin on the dorsum of the hand. The radial nerve itself innervates the brachioradialis and extensor carpi radialis longus. It then gives a deep branch that innervates the extensor carpi radialis brevis and the supinator before continuing as the posterior interosseous nerve, which innervates the extensor digitorum, extensor digiti minimi, extensor carpi ulnaris, abductor pollicis longus, extensor pollicis brevis, extensor pollicis longus, and extensor indicis *(Moore, pp 742, 761).*

53. **(E)** The flexor pollicis brevis is located medial to the abductor pollicis brevis. It flexes the thumb at the carpometacarpal and metacarpophalangeal joints and assists in opposition. Its tendon typically contains a sesamoid bone. It is innervated by the recurrent branch of the median nerve (C8–T1) *(Moore, p 767).*

54. **(E)** The palmaris brevis, innervated by the ulnar nerve, wrinkles the skin of the hypothenar eminence and deepens the hollow of the palm, assisting the palmar grip. The muscle actually covers and protects the ulnar artery and the ulnar nerve, which innervates it. The muscle is not by definition in the hypothenar compartment. The palmaris longus, on the other hand, flexes the hand at the wrist and tightens the palmar aponeurosis *(Moore, p 768).*

55. **(B)** The recurrent branch of the median nerve innervates the abductor pollicis brevis, flexor pollicis brevis, and opponens pollicis, but the deep branch of the ulnar nerve innervates adductor pollicis *(Moore, pp 769–770).*

56. **(C)** Lumbricals 1–2 and palmar interossei 1–3 are unipennate muscles. Lumbricals 3–4 and dorsal interossei 1–4 are bipennate muscles. The deltoid is multipennate *(Moore, p 770).*

57. **(C)** The deep branch of the ulnar nerve innervates the adductor pollicis, abductor digiti minimi, flexor digiti minimi brevis, opponens digiti minimi, lumbricals 3 and 4, dorsal interossei 1–4, and palmar interossei 1–3. The median nerve innervates lumbricals 1 and 2 *(Moore, p 770).*

58. **(E)** The carpal tunnel contains the median nerve, the four tendons of the flexor digitorum superficialis, the four tendons of the flexor digitorum profundus, and the tendon of the flexor pollicis longus *(Moore, p 774).*

59. **(A)** The sternoclavicular joint, which does not dislocate easily, is a saddle-type synovial joint but functions as a ball-and-socket joint. It is the articulation of the sternal end of the clavicle with the manubrium of the sternum. The joint is supplied by the internal thoracic and suprascapular arteries and is innervated by branches of the medial supraclavicular nerve and the nerve to the subclavius *(Moore, pp 781–782).*

60. **(E)** The acromioclavicular joint is a plane-type synovial joint and is strengthened by the AC ligament and the coracoclavicular ligament, which is composed of the conoid and trapezoid ligaments. It is supplied by the suprascapular and thoracoacromial arteries and is innervated by the supraclavicular, lateral pectoral, and axillary nerves. When dislocated, it is referred to as a "separated shoulder" *(Moore, pp 784, 787).*

61. **(B)** The pectoralis major (clavicular head) and deltoid (anterior part) flex the arm at the glenohumeral joint. The coracobrachialis and the biceps brachii assist *(Moore, p 792).*

62. **(A)** The posterior portion of the deltoid causes extension of the arm at the glenohumeral joint. The deltoid (as a whole, but especially the central part) causes abduction, whereas the pectoralis major and latissimus dorsi cause adduction. The subscapularis causes medial rotation, whereas the infraspinatus causes lateral rotation *(Moore, p 792).*

63. **(E)** The elbow is a hinge type of synovial joint, strengthened by radial and ulnar collateral ligaments. It is supplied by arteries derived from the anastomosis around the elbow and is innervated by the musculocutaneous, radial, and ulnar nerves. It is surrounded by the intratendinous olecranon bursa, the subtendinous olecranon bursa, and the subcutaneous olecranon bursa *(Moore, pp 795–798).*

64. **(C)** The proximal and distal radioulnar joints are pivot-type synovial joints. The radiocarpal (wrist) joint is a condyloid type of synovial joint. Intercarpal joints are plane-type synovial joints. Metacarpophalangeal joints are condyloid types of synovial joints. Interphalangeal joints are hinge-type synovial joints *(Moore, pp 800, 803, 807, 809).*

65. **(A)** All carpometacarpal and intermetacarpal joints are the plane-type synovial joints except for the carpometacarpal joint of the thumb, which is a saddle joint *(Moore, p 809).*

66. **(E)** The clavicle varies more in shape than most other long bones and is thicker and more curved in manual workers. The right clavicle is stronger than the left and is usually shorter. The clavicle can also be pierced by a branch of the supraclavicular nerve. The clavicle is a long bone with no medullary cavity. It consists of spongy (cancellous) bone with a shell of compact bone *(Moore, p 667).*

67. **(A)** Fractures of the scapula typically involve the protruding subcutaneous acromion. The remainder of the scapula is well protected by muscles and the thoracic wall itself *(Moore, p 669).*

68. **(D)** The surgical neck of the humerus is in direct contact with the axillary nerve, the radial nerve runs in the radial groove, the distal end of the humerus is in direct contact with the median nerve, and the medial epicondyle is in contact with the ulnar nerve *(Moore, p 670).*

69. **(A)** Damage to the long thoracic nerve results in "winging" of the scapula *(Moore, p 689).*

70. **(A)** The triangle of auscultation, a good place to examine lung sounds, is bounded by the superior horizontal border of the latissimus dorsi, the medial border of the scapula, and the inferolat-

eral border of the trapezius. The 6th and 7th ribs and the 6th intercostal space is subcutaneous *(Moore, p 693).*

71. **(B)** With paralysis of the latissimus dorsi, the patient is unable to raise the trunk as necessary for climbing. The cause could be injury to the thoracodorsal nerve (C6–C7–C8) *(Moore, p 693).*

72. **(A)** Injury to the dorsal scapular nerve (C4–C5) can paralyze the rhomboids, causing the scapula on one side to be located farther from the midline than that on the normal side *(Moore, p 695).*

73. **(B)** The deltoid atrophies when the axillary nerve (C5–C6) is damaged. Therefore, the rounded contour of the shoulder often disappears. A loss of sensation may occur on the lateral side of the proximal part of the arm *(Moore, pp 696–697).*

74. **(B)** Injury or disease may damage the rotator cuff, causing instability of the glenohumeral joint. The supraspinatus tendon is the most commonly torn part of the rotator cuff. Acute tears are uncommon in young persons *(Moore, pp 698–699).*

75. **(C)** In an upper brachial plexus injury causing Erb-Duchenne palsy, one would expect damage to C5–C6, resulting in "waiter's tip position" (adducted shoulder, medially rotated arm, and extended elbow). This results from paralysis of the deltoid, biceps, brachialis, and brachioradialis. The lateral aspect of the upper limb also experiences loss of sensation *(Moore, p 716).*

76. **(D)** Damage to the inferior trunks of the brachial plexus (C8–T1) affects the short muscles of the hand, resulting in "clawhand." The patient might have grabbed a tree limb to catch himself while falling to cause this injury. "Clawhand" may also be caused by an injury to the ulnar nerve *(Moore, pp 716–717, 761).*

77. **(B)** A knife wound to the axilla would damage the musculocutaneous nerve and result in paralysis of the coracobrachialis, biceps, and brachialis. Therefore, flexion of the elbow joint and supination of the forearm would be weakened. The patient would also lose sensation on the lateral surface of the forearm *(Moore, p 731).*

78. **(C)** When the median nerve is injured, the patient often exhibits the "hand of benediction." When the patient tries to make a fist, digits 2 and 3 remain partially extended because flexion of the PIP joints is lost in digits 1–3 and weakened in digits 4–5. Flexion of the DIP joints is lost in digits 2–3 but maintained in digits 4–5 (since the ulnar nerve controls the medial part of the flexor digitorum profundus). Flexion of the MCP joints of digits 2-3 will also be affected due to a loss of the lumbricals 1 and 2 *(Moore, pp 757, 774, 776).*

79. **(A)** The ulnar nerve is often injured where it passes posterior to the medial epicondyle of the humerus. The patient experiences loss of sensation in the medial part of the palm as well as in the medial 1½ digits. Most intrinsic hand muscles are paralyzed, and the patient loses the ability to adduct the hand at the wrist. Patients cannot make a fist since they are unable to flex the 4th and 5th digits at the DIP joints. The resulting deformity is known as "clawhand" *(Moore, pp 761, 776–777).*

80. **(E)** Amelia is the complete absence of one or more extremities while meromelia is the partial absence of one or more extremities. All segments of extremities are present but abnormally short in micromelia. In phocomelia, long bones are absent, and small hands or feet are attached to the trunk by short, irregular bones. In cleft hand (lobster claw deformity), the third metacarpal is absent and digits 1–2 and 4–5 are fused *(Sadler, p 179).*

81. **(C)** Syndactyly involves abnormal fusion of fingers and toes. Cleft hand (lobster claw deformity) consists of an abnormal cleft between the 2nd and 4th metacarpal bones, with the 3rd metacarpal and phalangeal bones being absent and with digits 1–2 and 4–5 being fused. Polydactyly involves extra fingers or toes, while ectrodactyly involves the absence of a digit. Mutations in *HOXA13* result in hand-foot-genital syndrome, where carpals and short digits are fused and the genitalia have altered structures *(Sadler, p 181).*

82. **(D)** During development, dorsal cells organize as the epimere and ventral cells organize as the hypomere. Dorsal rami innervate muscles derived from the epimere, whereas ventral rami innervate muscles derived from the hypomere. Myoblasts of the epimere form the extensor muscles of the vertebral column, and those of the hypomere give rise to muscles of the limbs and body wall. Somites and somitomeres form the musculature of the limbs *(Sadler, pp 189–190).*

83. **(E)** Partial or complete absence of one or more muscles is rather common. One of the best-known examples is total or partial absence of the pectoralis major (Poland anomaly). Similarly, the palmaris longus, serratus anterior, and quadratus femoris may be partially or entirely absent *(Sadler, p 192).*

84. biceps brachii

85. cephalic vein

86. radial nerve

87. brachial artery

88. median nerve

89. lateral cord

90. musculocutaneous nerve

91. medial antebrachial cutaneous nerve

92. ulnar nerve

93. lateral pectoral nerve

94. coracobrachialis

95. brachialis

96. ulna

97. long head of triceps brachii

98. lateral head of triceps brachii

CHAPTER 3

The Thorax

Questions

DIRECTIONS (Questions 1 through 75): Each of the numbered items or incomplete statements in this section is followed by answers or by completions of the statement. Select the ONE lettered answer or completion that is BEST in each case.

1. The articular part of a rib tubercle articulates with which of the following structures?

   (A) body of the vertebra
   (B) costal cartilage
   (C) adjacent rib
   (D) sternum
   (E) transverse process

2. Where is the groove for the subclavian artery located?

   (A) posterior to the scalene tubercle
   (B) on the clavicle
   (C) on the manubrium
   (D) at the sternal angle
   (E) at the angle of the 1st rib

3. The retromammary space is located between which of the following structures?

   (A) skin and the areola
   (B) pectoralis major and minor
   (C) breast and deep pectoral fascia
   (D) suspensory ligament and the skin
   (E) lactiferous sinus and the nipple

4. The high death rate associated with breast cancer is related to which of the following?

   (A) nerve supply
   (B) blood supply
   (C) venous drainage
   (D) poor imaging techniques
   (E) complex lymphatic drainage

5. Which of the following statements correctly apply to the internal thoracic artery?

   (A) It runs posterior to the transversus thoracis muscle.
   (B) It ends in the 6th intercostal space.
   (C) It divides into the superior and inferior epigastric arteries.
   (D) It runs posterior to the superior six ribs.
   (E) It gives rise to the posterior intercostal arteries.

6. Which of the following statements concerning the sternal angle is correct?

   (A) It lies at the level of the intervertebral disk between T4/T5.
   (B) It is flanked by the costal cartilage of the 3rd pair of costal cartilages.
   (C) It is crossed by the superior epigastric artery.
   (D) It lies in the epigastric fossa.
   (E) It is located 3 mm superior to the jugular notch.

7. The pleural cavity contains which of the following?

   (A) lungs
   (B) bronchi
   (C) serous pleural fluid
   (D) lymph nodes
   (E) pulmonary arteries and veins

8. The parietal pleura consists of all of the following parts EXCEPT

   (A) costal
   (B) pericardial
   (C) mediastinal
   (D) diaphragmatic
   (E) cervical

9. All of the following statements correctly apply to the right lung EXCEPT

   (A) The superior and oblique fissures divide it into three lobes.
   (B) It is larger and heavier than the left lung.
   (C) It is shorter and wider than the left lung.
   (D) It contains a thin, tonguelike process called the lingula.
   (E) It has three surfaces.

10. All of the following statements correctly apply to the left main bronchus EXCEPT

   (A) It is wider, shorter, and runs more vertically than the right main bronchus.
   (B) It passes anterior to the esophagus.
   (C) It passes anterior to the thoracic aorta.
   (D) It contains c-shaped rings of hyaline cartilage.
   (E) It arises at the level of the sternal angle.

11. All of the following statements correctly apply to a bronchopulmonary segment EXCEPT

   (A) It is separated from adjacent segments by connective tissue septa.
   (B) It is the largest subdivision of a lobe.
   (C) It is not resectable.
   (D) It is named according to the segmental bronchus supplying it.
   (E) It is a pyramid-shaped segment of the lung, with its apex facing the lung root and its base at the pleural surface.

12. Which of the following structures is located in the middle mediastinum?

   (A) thoracic duct
   (B) lungs
   (C) esophagus
   (D) heart
   (E) azygos vein

13. All of the following statements correctly apply to the right atrium EXCEPT

   (A) It receives blood from the superior and inferior vena cava and coronary sinus.
   (B) It forms the right side of the heart.
   (C) It contains the crista terminalis.
   (D) It contains the limbus fossae ovalis.
   (E) It contains trabeculae carneae.

14. The interventricular septum contains which of the following structures?

   (A) anterior papillary muscle
   (B) fossa ovalis
   (C) sinus venarum
   (D) sinoatrial node
   (E) conus arteriosus

15. All of the following statements correctly apply to the right coronary artery EXCEPT

   (A) Typically supplies the SA node in approximately 60% of people.
   (B) Typically supplies the AV node in approximately 80% of people.
   (C) Arises from the arch of the aorta.
   (D) Runs in the coronary sulcus.
   (E) Dominance is typical.

16. Which of the following is the basic structural unit for gas exchange in the lung?

   (A) terminal bronchioles
   (B) respiratory bronchioles
   (C) alveolar ducts
   (D) alveoli
   (E) bronchi

17. Which of the following structures carry highly oxygenated blood from the lungs to the heart?

   (A) pulmonary arteries
   (B) pulmonary veins
   (C) coronary arteries
   (D) cardiac veins
   (E) ascending aorta

18. Which of the following structures is located in the posterior mediastinum?

(A) lungs
(B) heart
(C) azygos vein
(D) superior vena cava
(E) right coronary artery

19. All of the following statements concerning spinal nerves are correct EXCEPT

(A) The dorsal and ventral rami are both motor and sensory.
(B) The cutaneous branches include anterior, lateral, and posterior branches.
(C) The dorsal root is both sensory and motor.
(D) The ventral root is pure motor.
(E) They supply a bandlike skin area known as a dermatome.

20. Which of the following statements correctly applies to the tricuspid valve?

(A) It guards the left atrioventricular orifice.
(B) It guards the conus arteriosus.
(C) It is also known as the mitral valve.
(D) Chordae tendineae attach to the free edges of the cusps.
(E) The apex of each cusp attaches to the fibrous ring around the orifice.

21. All of the following statements concerning the pericardium are correct EXCEPT

(A) The external layer of the sac is fibrous.
(B) The internal layer is reflected onto the heart as the visceral serous layer (epicardium).
(C) The internal layer of the fibrous sac is the partietal serous layer.
(D) The pericardial cavity is the potential space between the parietal and visceral serous layers.
(E) The fibrous pericardium is attached to the sternum by the pericardiacophrenic ligament.

22. All of the following veins drain into the coronary sinus EXCEPT

(A) anterior cardiac
(B) small cardiac
(C) middle cardiac
(D) great cardiac
(E) oblique vein of the left atrium

23. All of the following statements concerning the sinoatrial node are correct EXCEPT

(A) It is located near the superior end of the sulcus terminalis.
(B) It is located at the junction of the superior vena cava and the right atrium.
(C) It is known as the pacemaker of the heart.
(D) It is specialized cardiac muscle fiber.
(E) It is avascular.

24. Which of the following structures is located in the left ventricle?

(A) fossa ovalis
(B) crista terminalis
(C) opening of the coronary sinus
(D) conus arteriosus
(E) posterior papillary muscle

25. Which of the following structures is located in the left atrium?

(A) pectinate muscle
(B) pulmonary semilunar valves
(C) septomarginal trabeculae
(D) septal papillary muscle
(E) atrioventricular node

26. Which of the following is NOT part of the chest?

(A) 12 pairs of ribs
(B) sternum
(C) costal cartilages
(D) 12 thoracic vertebrae
(E) clavicle

27. Which of the following is NOT likely to contribute to chest pain?

(A) cardiac disease
(B) pulmonary disease
(C) thyroid disorders
(D) gallbladder disorders
(E) intestinal disorders

28. Which of the following is a special feature of a thoracic vertebra?

(A) foramen for vertebral artery
(B) dens for rotation
(C) short spinous processes
(D) costal facets on bodies
(E) costal facets on spinous processes

29. Which of the following is a saddle-type synovial joint?

(A) sternoclavicular joint
(B) manubriosternal joint
(C) interchondral joint
(D) intervertebral joint
(E) costochondral joint

30. The mammary glands are modified versions of which type of gland?

(A) sebaceous gland
(B) lymph gland
(C) sweat gland
(D) tonsillar tissue
(E) endocrine gland

31. Which of the following does NOT supply the breast with blood?

(A) lateral thoracic artery
(B) thoracoacromial artery
(C) posterior intercostals arteries
(D) internal thoracic artery
(E) costoclavicular artery

32. Which of the following are innervated by dorsal rami?

(A) levatores costarum
(B) external intercostals
(C) internal intercostals
(D) subcostals
(E) serratus posterior inferior

33. Which of the following do NOT elevate the ribs?

(A) serratus posterior superior
(B) serratus posterior inferior
(C) external intercostals
(D) levatores costarum
(E) subcostals

34. With which of the following does the intercostobrachial nerve communicate?

(A) medial brachial cutaneous nerve
(B) thoracodorsal nerve
(C) long thoracic nerve
(D) first intercostal nerve
(E) sympathetic trunk

35. The musculophrenic arteries give rise to which of the following?

(A) anterior intercostal arteries for intercostal spaces 7–9
(B) posterior intercostals arteries for intercostal spaces 3–11
(C) subcostal artery
(D) inferior phrenic artery
(E) lumbar arteries

36. Anterior and posterior intercostals arteries pass between which two layers?

(A) skin and external intercostal muscles
(B) external and internal intercostal muscles
(C) internal and innermost intercostals muscles
(D) innermost intercostal muscles and transversalis fascia
(E) transversalis fascia and peritoneum

37. Of the bronchopulmonary segments in the superior lobe of the left lung, which two are combined to form the lingula?

(A) apical and posterior
(B) posterior and anterior

(C) anterior and superior
(D) superior and inferior
(E) apical and inferior

38. Which of the following areas do NOT receive visceral afferent fibers?

(A) bronchial mucosa
(B) bronchial muscles
(C) interalveolar connective tissue
(D) pulmonary arteries and veins
(E) diaphragm

39. Sympathetic nerves do NOT contribute to which of the following?

(A) diaphragm
(B) bronchial muscle
(C) pulmonary vessels
(D) alveolar glands of the bronchial tree
(E) sweat glands of the chest

40. Which of the following is correct?

(A) While a person is supine, the arch of the aorta lies superior to the transverse thoracic plane.
(B) While a person is supine, the bifurcation of the trachea lies at the level of the xiphisternal junction and T9.
(C) While a person is supine, the central tendon of the diaphragm is transected by the transverse thoracic plane.
(D) While a person is standing, the arch of the aorta lies inferior to the transverse thoracic plane.
(E) While a person is standing, the tracheal bifurcation is transected by the transverse thoracic plane.

41. The pericardium does NOT receive blood supply from which of the following?

(A) pericardiacophrenic artery
(B) musculophrenic artery
(C) bronchial arteries
(D) esophageal arteries
(E) pectoral arteries

42. Which of the following is NOT a branch of the right coronary artery?

(A) SA nodal
(B) right marginal
(C) posterior interventricular
(D) AV nodal
(E) circumflex

43. Which of the following is NOT contained in the superior mediastinum?

(A) thymus
(B) great vessels and their branches
(C) azygos and hemiazygos veins
(D) trachea
(E) esophagus

44. The thymus is supplied by which of the following?

(A) anterior intercostal branches of the internal thoracic artery
(B) brachiocephalic artery
(C) posterior intercostal arteries
(D) left common carotid artery
(E) pericardiacophrenic arteries

45. Which of the following is NOT a branch of the aorta in the thorax?

(A) posterior intercostal arteries
(B) inferior phrenic artery
(C) bronchial arteries
(D) esophageal arteries
(E) superior phrenic arteries

46. The posterior mediastinum does NOT include which of the following?

(A) prevertebral muscles
(B) thoracic aorta
(C) thoracic duct
(D) thoracic sympathetic trunks
(E) thoracic splanchnic nerves

47. Which of the following may compress the esophagus?

(A) aortic arch
(B) right main bronchus
(C) left vagus
(D) right brachiocephalic vein
(E) brachiocephalic trunk

48. The azygos vein receives blood from which of the following?

(A) posterior intercostal veins
(B) anterior intercostal veins
(C) left internal jugular vein
(D) anterior jugular vein
(E) lateral thoracic vein

49. The hemiazygos vein does NOT receive blood from which of the following?

(A) left subcostal vein
(B) ascending lumbar veins
(C) inferior three posterior intercostal veins
(D) superior phrenic veins
(E) small mediastinal veins

50. The accessory hemiazygos vein parallels the vertebral column along which vertebral levels?

(A) T1–4
(B) T2–6
(C) T5–8
(D) T8–12
(E) L1–4

51. Which of the following is NOT contained in the anterior mediastinum?

(A) fat
(B) sternopericardial ligaments
(C) branches of internal thoracic vessels
(D) lymphatic vessels
(E) heart

52. Which of the following is NOT part of the sympathetic trunks?

(A) prevertebral ganglia
(B) paravertebral ganglia
(C) ganglion impar
(D) superior cervical ganglia
(E) inferior cervical ganglia

53. Which of the following does NOT contain synapses for the sympathetic nervous system?

(A) prevertebral ganglia
(B) collateral ganglia
(C) aortic plexus
(D) celiac ganglia
(E) cardiac plexus

54. Which of the following does NOT occur in the sympathetic trunk?

(A) Presynaptic neurons synapse with postsynaptic neurons immediately.
(B) Fibers ascend and synapse.
(C) Fibers descend and synapse.
(D) Presynaptic fibers innervate immediately surrounding blood vessels.
(E) Fibers pass, without synapsing, into a splanchnic nerve.

55. Parasympathetic fibers stimulate secretion by all glands except which of the following?

(A) sweat glands
(B) lacrimal glands
(C) salivary glands
(D) nasal glands
(E) palatine glands

56. Vasoconstriction is sympathetically stimulated with the exception of which arteries?

(A) bronchial arteries
(B) coronary arteries
(C) esophageal arteries
(D) adrenal arteries
(E) renal arteries

57. Which of the following is a result of sympathetic stimulation in the heart?

(A) decrease in the rate and strength of contraction

(B) inhibition of the effect of parasympathetic system on coronary arteries, allowing them to dilate
(C) production of atrial natriuretic factor
(D) opening and closing of mitral valve
(E) glandular secretion

58. Which of the following is NOT a result of sympathetic stimulation in the lungs?

(A) inhibition of parasympathetic system
(B) bronchodilation
(C) reduced secretion
(D) maximum air exchange
(E) surfactant production

59. Sympathetic fibers of the heart are accompanied by visceral afferent fibers that convey which type of sensation?

(A) reflex
(B) pain
(C) rate information
(D) contraction feedback
(E) pressure sensation

60. The pulmonary plexus contains which of the following?

(A) cell bodies of visceral afferent fibers
(B) cell bodies of postsynaptic sympathetic fibers
(C) cell bodies of postsynaptic parasympathetic fibers
(D) somatic efferent fibers passing on to diaphragm
(E) somatic afferent fibers from diaphragm

61. Visceral afferent fibers of the tenth cranial nerve are NOT distributed to which of the following?

(A) bronchial mucosa
(B) bronchial muscles
(C) interalveolar connective tissue
(D) pulmonary arteries and veins
(E) diaphragm

62. The greater, lesser, and least splanchnic nerves are examples of what type of splanchnic nerves?

(A) cervical splanchnic nerves
(B) upper thoracic splanchnic nerves
(C) lower thoracic splanchnic nerves
(D) lumbar splanchnic nerves
(E) pelvic splanchnic nerves

63. The greater splanchnic nerve originates from which vertebral levels?

(A) T1–4
(B) T5–9
(C) T10–11
(D) T12
(E) L1–4

64. The fibers of the greater splanchnic nerve synapse in which ganglion?

(A) celiac ganglion
(B) inferior mesenteric ganglion
(C) paravertebral ganglia
(D) pulmonary plexus
(E) adrenal cortex

65. The level of the domes of the diaphragm varies according to various situations. Which of the following does NOT affect the level of the domes of the diaphragm?

(A) phase of respiration
(B) posture
(C) size of distention of abdominal viscera
(D) degree of distention of abdominal viscera
(E) heart contractions

66. Which of the following passes through the caval opening of the diaphragm?

(A) terminal branches of the right phrenic nerve
(B) thoracic duct
(C) greater thoracic splanchnic nerve
(D) sympathetic trunk
(E) vagus

67. Which of the following does NOT pass through the esophageal hiatus of the diaphragm?

(A) esophagus
(B) branches of the left gastric vessels
(C) lymphatic vessels
(D) vagal trunks
(E) lesser thoracic splanchnic nerve

68. Which of the following passes through the aortic hiatus of the diaphragm?

(A) least thoracic splanchnic nerve
(B) thoracic duct
(C) branches of right gastric vessels
(D) sympathetic trunks
(E) terminal branches of the left phrenic nerve

69. Ribs are formed from which of the following?

(A) sclerotome portion of paraxial mesoderm
(B) lateral plate mesoderm
(C) ectodermal invagination
(D) endodermal migration
(E) neural crest cell transitory development

70. The pleuropericardial membranes develop into what structures of the adult?

(A) fibrous pericardium
(B) diaphragm
(C) parietal pleura
(D) visceral pleura
(E) the membranes degenerate completely

71. The diaphragm is NOT derived from which of the following?

(A) septum transversum
(B) pleuroperitoneal membranes
(C) muscular components from lateral and dorsal body walls
(D) mesentery of esophagus
(E) anterior thoracic fascia

72. The entire cardiovascular system is derived from which of the following?

(A) neural crest cells
(B) endoderm
(C) mesoderm
(D) ectoderm
(E) notochord invagination

73. Which of the following is NOT derived from the original aortic arch system?

(A) carotid arteries
(B) arch of the aorta
(C) pulmonary artery
(D) right subclavian artery
(E) coronary arteries

74. The respiratory system is an outgrowth of what?

(A) middle mediastinum
(B) ventral wall of foregut
(C) anterior abdominal wall
(D) aortic arches
(E) pharyngeal arches

75. Which of the following correctly describes the development of the lungs, in order?

(A) pseudoglandular period, canalicular period, terminal sac period, alveolar period
(B) canalicular period, pseudoglandular period, terminal sac period, alveolar period
(C) alveolar period, pseudoglandular period, canalicular period, terminal sac period
(D) pseudoglandular period, terminal sac period, alveolar period, canalicular period
(E) terminal sac period, alveolar period, pseudoglandular period, canalicular period

DIRECTIONS (Questions 76 through 85): Identify the anatomical features indicated on the art below.

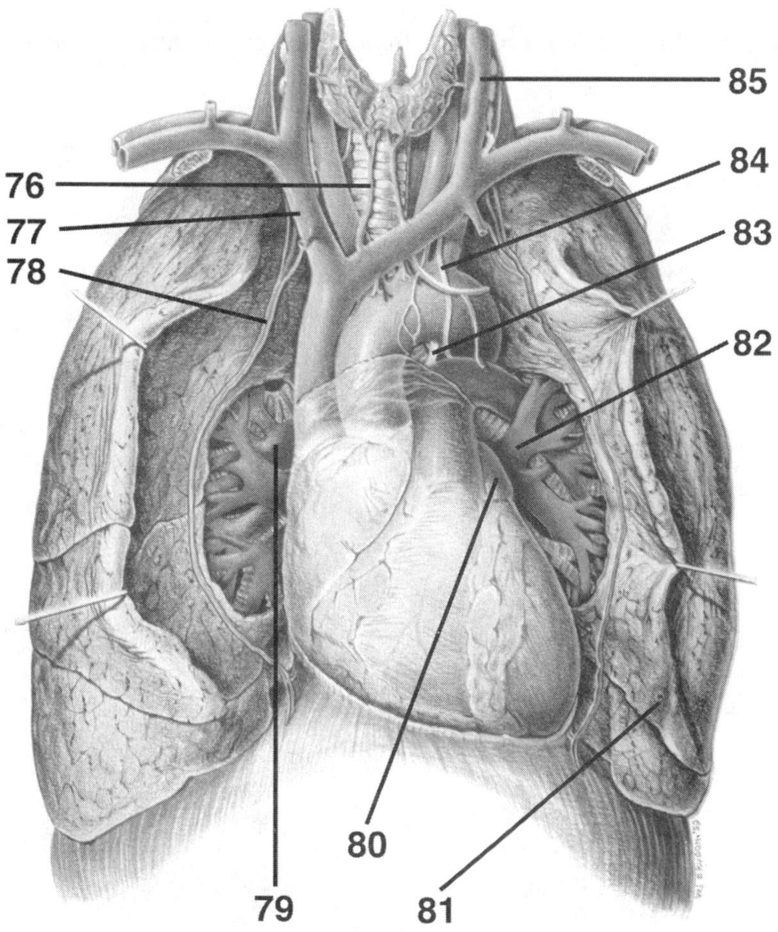

DIRECTIONS (Questions 86 through 90): Identify the anatomical features indicated on the art below.

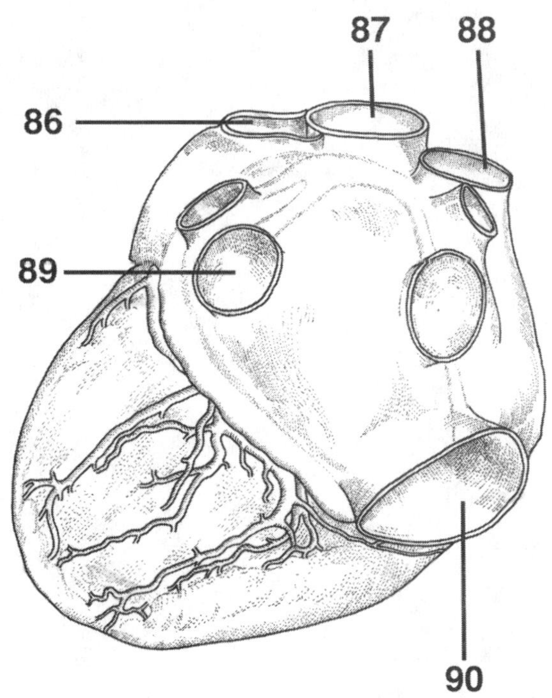

DIRECTIONS (Questions 91 through 95): Identify the anatomical features indicated on the art below.

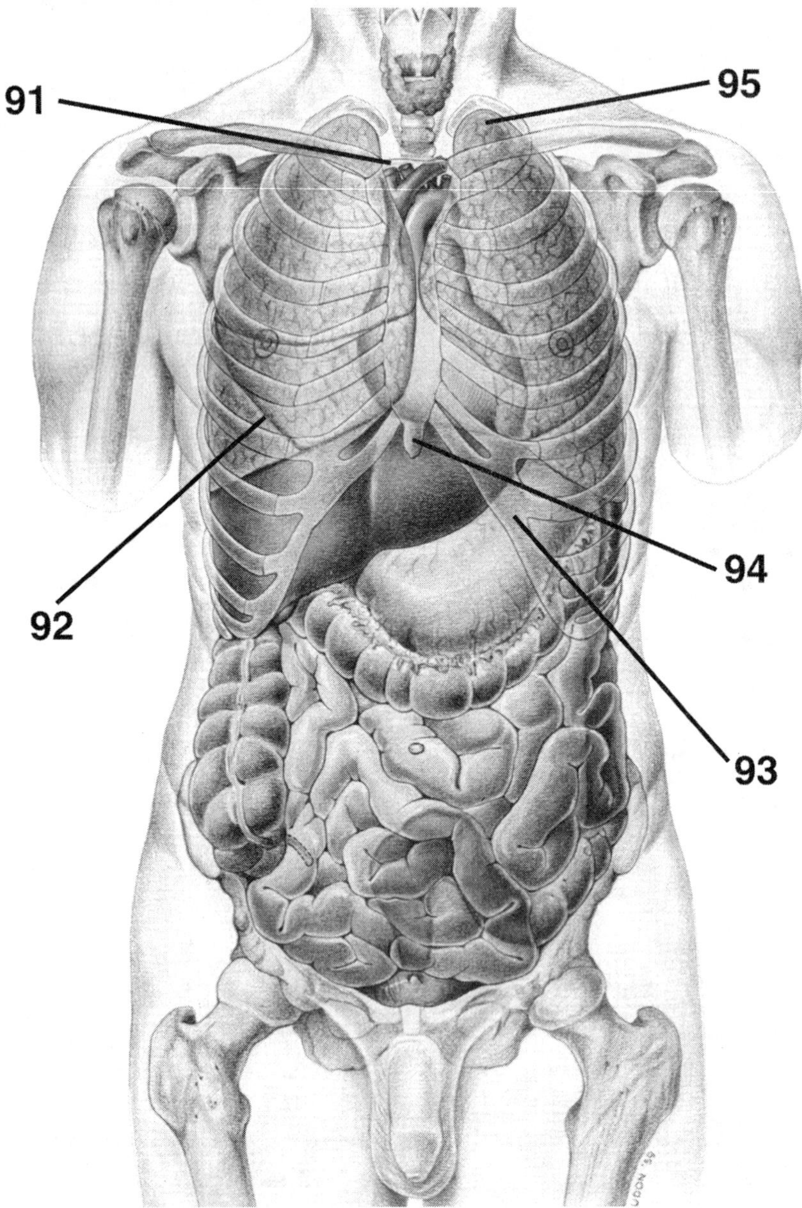

Answers and Explanations

1. **(E)** The tubercle has a smooth articular part for articulating with the corresponding transverse process of the vertebra and a rough nonarticular part for attachment of the costotransverse ligament *(Moore, p 63)*.

2. **(A)** The surface of the 1st rib has two transversely directed shallow grooves, anterior and posterior to the scalene tubercle, for the subclavian vein and the subclavian artery respectively *(Moore, pp 63–64)*.

3. **(C)** Between the breast and deep pectoral fascia is a loose connective tissue plane or potential space known as the retromammary space *(Moore, p 73)*.

4. **(E)** Because the axillary lymph nodes are the most common site of metastases from a breast cancer, enlargement of the palpable nodes in a woman suggests the possibility of breast cancer and may be key to early detection. However, the absence of enlarged axillary nodes is no guarantee that metastasis from a breast cancer has not occurred, because the malignant cells may have passed to other nodes, such as the infraclavicular and supraclavicular lymph nodes *(Moore, p 78)*.

5. **(B)** After descending past the 2nd costal cartilage, the internal thoracic artery runs anterior to the transversus thoracis muscle. It ends in the 6th intercostal space, where it divides into the superior epigastric and musculophrenic arteries *(Moore, p 91)*.

6. **(A)** The sternal angle is located opposite the 2nd pair of costal cartilages at the level of the 4th intervertebral disc between T4 and T5 vertebrae *(Moore, p 66)*.

7. **(C)** The pleural cavity, the potential space between the layers of pleura, contains a capillary layer of serous pleural fluid. It provides the lubrication and cohesion that keep the lung surface in contact with the thoracic wall *(Moore, p 95)*.

8. **(B)** The parietal pleura includes costal, mediastinal, diaphragmatic, and cervical parts *(Moore, pp 95–96)*.

9. **(D)** The right lung has three lobes, the left two. The right lung is larger and heavier than the left, but it is shorter and wider, because the right dome of the diaphragm is higher and the heart and pericardium bulge more to the left. The anterior margin of the right lung is relatively straight, whereas the margin of the left lung has a deep cardiac notch. The cardiac notch primarily indents the anteroinferior aspect of the superior lobe of the left lung. This often creates a thin, tongue-like process of the superior lobe, called the lingula *(Moore, p 101)*.

10. **(A)** The right main bronchus is wider and shorter, running more vertically than the left main bronchus as it passes directly to the hilum of the lung. The left main bronchus passes inferolaterally, inferior to the arch of the aorta and anterior to the esophagus and thoracic aorta, to reach the hilum of the lung *(Moore, p 104)*.

11. **(C)** A bronchopulmonary segment is a pyramid-shaped segment of the lung with its apex facing the lung root and its base at the pleural surface. It is the largest subdivision of a lobe and is sepa-

rated from adjacent segments by connective tissue. It is surgically resectable *(Moore, p 104).*

12. **(D)** The middle mediastinum contains the heart *(Moore, p 114).*

13. **(E)** The right border of the heart is formed by the right atrium; it receives venous blood from the SVC, IVC, and coronary sinus. It contains both the crista terminalis and the limbus fossae ovalis. The interior of the ventricles contains irregular muscular elevations called trabeculae carneae *(Moore, pp 125–127).*

14. **(E)** The interventricular septum is composed of membranous and muscular parts. The conus arteriosus leads into the pulmonary trunk. A thick muscular ridge, the supraventricular crest, separates the ridged muscular wall of the inflow part of the right ventricle from the smooth wall of the conus arteriosus or outflow part *(Moore, p 127).*

15. **(C)** Typically, the right coronary artery supplies the right atrium, most of the right ventricle, part of the left ventricle, part of the AV septum, and the SA node in approximately 60% of people and the AV node in approximately 80% of people *(Moore, p 135).*

16. **(D)** The alveolus is the basic unit of gas exchange in the lung *(Moore, p 104).*

17. **(B)** The pulmonary veins, two on each side, carry well-oxygenated ("arterial") blood from the lungs to the left atrium of the heart *(Moore, p 107).*

18. **(C)** The posterior mediastinum contains the thoracic aorta, thoracic duct, azygos and hemiazygos veins, esophagus, and thoracic sympathetic trunks *(Moore, pp 150–151).*

19. **(C)** The dorsal and ventral rami are both motor and sensory. The cutaneous branches include anterior, lateral, and posterior branches. The dorsal root is sensory and the ventral root is motor. The spinal nerves supply a band-like skin area known as a dermatome *(Moore, p 85).*

20. **(D)** The tricuspid valve guards the right AV orifice. The bases of the valve cusps are attached to the fibrous ring around the orifice. Chordae tendineae attach to the free edges and ventricular surfaces of the anterior, posterior, and septal cusps *(Moore, p 127).*

21. **(E)** The fibrous pericardium is attached to the posterior surface of the sternum by the sternopericardial ligaments *(Moore, p 116).*

22. **(A)** The coronary sinus receives the great and small cardiac veins, middle cardiac vein, left posterior ventricular vein, and left marginal vein. The anterior cardiac veins begin over the anterior surface of the right ventricle and cross over the coronary groove to end directly in the right atrium *(Moore, pp 136–137).*

23. **(E)** The SA node is located anterolaterally just deep to the epicardium at the junction of the SVC and right atrium, near the superior end of the sulcus terminalis. The SA node, a small collection of nodal tissue and specialized cardiac muscle fibers, is the pacemaker of the heart. The SA node is supplied by a branch of the right coronary artery in about 60% of individuals *(Moore, p 137).*

24. **(E)** The interior of the left ventricle includes anterior and posterior papillary muscles *(Moore, p 131).*

25. **(A)** The interior of the left atrium has a large-walled part and a smaller muscular auricle containing pectinate muscles *(Moore, p 129).*

26. **(E)** The chest consists of 12 pairs of ribs, the sternum, costal cartilages, and 12 thoracic vertebrae *(Moore, p 60).*

27. **(C)** Chest pain may result from cardiac disease, pulmonary disease, intestinal problems, gallbladder disorders, and musculoskeletal disorders *(Moore, p 61).*

28. **(D)** Thoracic vertebrae have costal facets or demifacets on their bodies, costal facets on the transverse processes (for the first 9 or 10 thoracic vertebrae), and long spinous processes *(Moore, p 65).*

29. **(A)** The sternoclavicular joint is a saddle-type synovial joint *(Moore, p 69).*

30. **(C)** The mammary glands are modified sweat glands *(Moore, p 74).*

31. **(E)** The breast is supplied by the internal thoracic, lateral thoracic, thoracoacromial, and posterior intercostal arteries *(Moore, p 75).*

32. **(A)** The levatores costarum are innervated by the dorsal primary rami of C8–T11. *(Moore, p 84).*

33. **(B)** The serratus posterior inferior depresses the ribs *(Moore, p 84).*

34. **(A)** The intercostobrachial nerve communicates with the medial brachial cutaneous nerve *(Moore, p 87).*

35. **(A)** The musculophrenic artery gives rise to anterior intercostal arteries for intercostal spaces 7–9 *(Moore, p 88).*

36. **(C)** Anterior and posterior intercostal arteries pass between the internal and innermost intercostal muscles *(Moore, p 88).*

37. **(D)** The superior and inferior bronchopulmonary segments of the superior lobe of the left lung combine to form the lingula *(Moore, p 106).*

38. **(E)** Visceral afferent fibers are distributed to bronchial mucosa, bronchial muscles, interalveolar connective tissue, and pulmonary arteries and veins *(Moore, p 111).*

39. **(A)** The diaphragm receives somatic efferent and afferent innervation—not sympathetic innervation *(Moore, p 113).*

40. **(A)** While a person is supine, the arch of the aorta lies superior to the transverse thoracic plane *(Moore, p 115).*

41. **(E)** The pericardium receives blood supply from the pericardiacophrenic artery, musculophrenic artery, bronchial arteries, esophageal arteries, superior phrenic arteries, and coronary arteries *(Moore, p 120).*

42. **(E)** The branches of the right coronary artery include the SA nodal, right marginal, posterior interventricular, and AV nodal arteries *(Moore, p 134).*

43. **(C)** The superior mediastinum contains the thymus, great vessels and their branches, vagus nerves, phrenic nerves, cardiac plexus, left recurrent laryngeal nerve, trachea, esophagus, thoracic duct, and prevertebral muscles *(Moore, p 142).*

44. **(A)** The thymus is supplied by the anterior intercostal and anterior mediastinal branches of the internal thoracic artery *(Moore, p 142).*

45. **(B)** Branches of the aorta in the thorax include the posterior intercostal, bronchial, esophageal, and superior phrenic arteries *(Moore, p 145).*

46. **(A)** The posterior mediastinum includes the thoracic aorta, thoracic duct, posterior mediastinal lymph nodes, azygos and hemiazygos veins, esophagus, esophageal plexus, thoracic sympathetic trunks, and thoracic splanchnic nerves *(Moore, pp 150–151).*

47. **(A)** The esophagus is compressed by the aortic arch, left main bronchus, and diaphragm *(Moore, p 152).*

48. **(A)** The azygos vein receives blood from the posterior intercostal veins, vertebral venous plexuses, mediastinal veins, esophageal veins, and bronchial veins *(Moore, p 155).*

49. **(D)** The hemiazygos vein receives blood from the left subcostal vein, ascending lumbar veins, the inferior three posterior intercostal veins, inferior esophageal veins, and small mediastinal veins *(Moore, p 155).*

50. **(C)** The accessory hemiazygos descends on the left side of the vertebral column from T5 through T8 *(Moore, p 155).*

51. **(E)** The anterior mediastinum contains fat, sternopericardial ligaments, branches of internal thoracic vessels, and lymphatic vessels *(Moore, p 156).*

52. **(A)** The sympathetic trunk contains the superior, middle, and inferior cervical ganglia; paravertebral ganglia; and the ganglion impar *(Moore, p 46).*

53. **(E)** The cardiac plexus does not contain synapses for sympathetic fibers, as these fibers have already synapsed in the sympathetic trunk *(Moore, p 150).*

54. **(D)** In the synaptic trunk, presynaptic neurons synapse with postsynaptic neurons immediately. Some fibers ascend and synapse, while other fibers descend and synapse. Some fibers pass into a splanchnic nerve without synapsing *(Moore, p 46).*

55. **(A)** With the exception of sweat glands, glandular secretion is parasympathetically stimulated *(Moore, p 51).*

56. **(B)** With the exception of coronary arteries, vasoconstriction is sympathetically stimulated *(Moore, p 51).*

57. **(B)** Sympathetic stimulation of the heart increases the rate and strength of contraction and inhibits the effect of the parasympathetic system on coronary arteries, allowing them to dilate *(Moore, p 51).*

58. **(E)** Sympathetic stimulation of the lungs results in inhibition of the parasympathetic system, bronchodilation, reduced secretion, and maximum air exchange *(Moore, p 51).*

59. **(B)** Sympathetic fibers of the heart are accompanied by visceral afferent fibers, which convey pain information *(Moore, p 52).*

60. **(C)** The pulmonary plexus contains cell bodies of postsynaptic parasympathetic fibers *(Moore, p 110).*

61. **(E)** Visceral afferent fibers of the tenth cranial nerve are distributed to bronchial mucosa (for cough reflex), bronchial muscles (for stretch reception), interalveolar connective tissue (for Hering-Breuer reflexes, limiting respiratory excursions), and pulmonary arteries (as pressure receptors) and veins (as chemoreceptors) *(Moore, p 111).*

62. **(C)** The greater, lesser, and least splanchnic nerves are lower thoracic splanchnic nerves *(Moore, p 301).*

63. **(B)** The greater thoracic splanchnic nerve originates from vertebral levels T5–9 *(Moore, p 301).*

64. **(A)** The fibers of the greater splanchnic nerve synapse in the celiac ganglion *(Moore, p 302).*

65. **(E)** The level of the domes of the diaphragm varies according to phase of respiration, posture, and size and degree of distention of abdominal viscera *(Moore, p 289).*

66. **(A)** The inferior vena cava, terminal branches of the right phrenic nerve, and lymphatics pass through the caval opening of the diaphragm *(Moore, p 294).*

67. **(E)** The esophagus, anterior and posterior vagal trunks, esophageal branches of the left gastric vessels, and lymphatic vessels pass through the esophageal hiatus *(Moore, p 294).*

68. **(B)** The aorta, thoracic duct, and azygos vein (sometimes) pass through the aortic hiatus *(Moore, p 295).*

69. **(A)** Ribs form from costal processes of thoracic vertebrae and thus are derived from the sclerotome of paraxial medoderm *(Sadler, p 184).*

70. **(A)** The pleuropericardial membranes develop into the fibrous pericardium *(Sadler, p 201).*

71. **(E)** The diaphragm is derived from the septum transversum and pleuroperitoneal membranes. Muscular components stem from lateral and dorsal body walls and mesentery of the esophagus *(Sadler, p 206).*

72. **(C)** The entire cardiovascular system is derived from mesoderm. *(Sadler, p 256)*

73. **(E)** The carotid arteries, arch of the aorta, pulmonary artery, and right subclavian artery are

derived from the original aortic arch system *(Sadler, p 257).*

74. **(B)** The respiratory system is an outgrowth of the ventral wall of the foregut *(Sadler, p 268)*

75. **(A)** The lungs develop in the following periods: pseudoglandular period, canalicular period, terminal sac period, and alveolar period *(Sadler, p 266).*

76. inferior thyroid vein

77. right brachiocephalic vein

78. phrenic nerve and pericardiacophrenic vessels

79. right pulmonary artery

80. left auricle

81. lingula

82. left pulmonary vein

83. ligamentum arteriosum

84. left vagus nerve

85. left internal jugular vein

86. pulmonary arteries

87. aorta

88. superior vena cava

89. left inferior pulmonary vein

90. inferior vena cava

91. manubrium of sternum

92. oblique fissure

93. costal cartilage

94. xiphoid process

95. copula

CHAPTER 4

The Abdomen

Questions

DIRECTIONS (Questions 1 through 100): Each of the numbered items or incomplete statements in this section is followed by answers or by completions of the statement. Select the ONE lettered answer or completion that is BEST in each case.

1. For general clinical descriptions, which of the following planes is used as one of the planes to define four quadrants of the abdominal cavity?

   (A) subcostal
   (B) transtubercular
   (C) midclavicular
   (D) transumbilical
   (E) midaxillary

2. The midclavicular planes pass through the midpoint of the clavicles to the midpoint of which of the following structures?

   (A) anterior superior iliac spine
   (B) symphysis pubis
   (C) umbilicus
   (D) inguinal ligament
   (E) xiphoid process

3. The fascial layer that covers the deep surface of the transverse abdominal muscle is known as which of the following?

   (A) parietal peritoneum
   (B) deep fascia
   (C) transversalis fascia
   (D) Scarpa's fascia
   (E) Camper's fascia

4. Where is extraperitoneal fat located?

   (A) between the abdominal oblique muscles
   (B) deep to the parietal peritoneum
   (C) superficial to Camper's fascia
   (D) superficial to the deep fascia
   (E) deep to the transversalis fascia

5. The superficial muscle fibers of the external abdominal oblique arising from the middle to lower ribs interdigitate with which of the following muscles?

   (A) internal abdominal oblique
   (B) serratus anterior
   (C) rectus abdominis
   (D) pyramidalis
   (E) transversus abdominis

6. The anterolateral abdominal wall is bounded by all of the following structures EXCEPT

   (A) cartilages of the 7th through 10th ribs
   (B) linea alba
   (C) xiphoid process
   (D) inguinal ligament
   (E) pelvic bone

7. The aponeuroses of all three flat muscles of the anterolateral abdominal wall interlace in which of the following structures?

   (A) inguinal ligament
   (B) transversalis fascia
   (C) linea alba
   (D) anterior superior iliac spine
   (E) rectus abdominis

8. All of the following structures are located within the rectus sheath EXCEPT

(A) pyramidalis
(B) rectus abdominis
(C) inferior epigastric arteries and veins
(D) deep inguinal ring
(E) ventral primary rami of T7–T12 nerves

9. Inferiorly, the inferior margin of the external oblique aponeurosis thickens and folds back on itself to form which of the following structures?

(A) rectus sheath
(B) inguinal ligament
(C) arcuate line
(D) deep inguinal ring
(E) fundiform ligament

10. The reflected inguinal ligament receives fibers from the contralateral aponeurosis of which of the following structures?

(A) external abdominal oblique
(B) internal abdominal oblique
(C) transverse abdominal oblique
(D) rectus abdominis
(E) pyramidalis

11. Between the internal oblique and transverse abdominal muscles is a neurovascular plane that contains all of the following EXCEPT

(A) iliohypogastric nerve
(B) deep circumflex iliac artery
(C) inferior epigastric artery
(D) subcostal nerve
(E) ilioinguinal nerve

12. The rectus abdominis muscle is anchored transversely by attachment to the anterior layer of the rectus sheath by which of the following structures?

(A) pubic tubercle
(B) xiphoid process
(C) linea alba
(D) tendinous intersections
(E) umbilicus

13. Which of the following structures defines the point at which the posterior lamina of the internal oblique and the aponeurosis of the transverse abdominal become part of the anterior rectus sheath?

(A) arcuate line
(B) inguinal ligament
(C) tendinous intersections
(D) deep inguinal ring
(E) medial crus

14. The two medial umbilical folds represent remnants of which of the following structures?

(A) urachus
(B) umbilical arteries
(C) umbilical veins
(D) ductus venosus
(E) ductus arteriosus

15. Which of the following fossae are potential sites for direct inguinal hernias?

(A) supravesical
(B) medial inguinal
(C) lateral inguinal
(D) ischiorectal
(E) iliac

16. The inguinal canal contains which of the following nerves?

(A) iliohypogastric
(B) ilioinguinal
(C) genital branch of the genitofemoral
(D) obturator
(E) lateral femoral cutaneous

17. Which of the following structures give rise to the deep inguinal ring?

(A) gubernaculum
(B) conjoined tendon
(C) lacunar ligament
(D) external abdominal oblique aponeurosis
(E) transversalis fascia

18. The lacunar ligament is a reflection or extension from the deep aspect of which of the following structures?

(A) falciform ligament
(B) round ligament
(C) rectus sheath
(D) inguinal ligament
(E) transversalis fascia

19. The iliopubic tract is the thickened inferior margin of which of the following structures?

(A) inguinal ligament
(B) transversalis fascia
(C) conjoined tendon
(D) falciform ligament
(E) round ligament

20. The testes develop in which of the following areas?

(A) scrotum
(B) abdominal cavity
(C) extraperitoneal
(D) rectus sheath
(E) superficial fascia

21. The gubernaculum is represented postnatally by which of the following structures?

(A) tunica vaginalis testes
(B) processus vaginalis
(C) ductus deferens
(D) scrotal ligament
(E) internal spermatic fascia

22. The cremaster muscle and fascia are derived from which of the following structures?

(A) external abdominal oblique muscle
(B) transverse abdominal muscle and fascia
(C) internal abdominal muscle
(D) transversalis fascia
(E) external abdominal aponeurosis

23. The cremaster muscle is innervated by which of the following nerves?

(A) genital branch of the genitofemoral
(B) ilioinguinal
(C) T12
(D) femoral
(E) obturator

24. The artery of the ductus deferens arises from which of the following structures?

(A) aorta
(B) inferior epigastric
(C) inferior vesical
(D) pudendal
(E) common iliac

25. All of the following nerves contribute branches to the scrotum EXCEPT

(A) lateral femoral cutaneous
(B) pudendal
(C) ilioinguinal
(D) genitofemoral
(E) posterior femoral cutaneous

26. The epididymis is located on the posterior aspect of which of the following structures?

(A) urinary bladder
(B) prostate
(C) testis
(D) ovary
(E) uterine tube

27. The testes are covered by a tough fibrous coat known as the

(A) cremaster fascia
(B) tunica albuginea
(C) gubernaculum
(D) tunica dartos
(E) Scarpa's fascia

28. Which of the following structures represents the closed-off distal part of the embryonic processus vaginalis?

(A) gubernaculum
(B) tunica albuginea
(C) epididymis
(D) tunica vaginalis
(E) urogenital diaphragm

29. The pampiniform plexus provides which of the following testicular functions?

(A) lymphatic drainage
(B) blood supply
(C) nerve supply
(D) thermoregulatory system
(E) hormonal production

30. The parasympathetic innervation of the testis includes which of the following nerves?

(A) pelvic splanchnic
(B) vagus
(C) iliohypogastric
(D) ilioinguinal
(E) pudendal

31. The peritoneal cavity contains which of the following?

(A) liver
(B) pancreas
(C) large intestine
(D) kidney
(E) peritoneal fluid

32. Which of the following structures connects the lesser curvature of the stomach and the proximal part of the duodenum to the liver?

(A) lesser omentum
(B) peritoneal ligament
(C) mesentery
(D) lesser omentum
(E) peritoneal fold

33. Which of the following structures is often referred to as the "abdominal policeman"?

(A) hepatoduodenal ligament
(B) gastrohepatic ligament
(C) greater omentum
(D) gastrocolic ligament
(E) falciform ligament

34. Which of the following ligaments conducts the portal triad (portal vein, hepatic artery, and bile duct)?

(A) greater omentum
(B) falciform ligament
(C) gastrohepatic ligament
(D) hepatoduodenal ligament
(E) gastrocolic ligament

35. Which of the following structures forms the superior boundary of the omental foramen?

(A) inferior vena cava
(B) duodenum
(C) caudate lobe of the liver
(D) head of the pancreas
(E) hepatoduodenal ligament

36. Which of the following structures contains both smooth and skeletal muscles?

(A) stomach
(B) jejunum
(C) cecum
(D) esophagus
(E) rectum

37. Which of the following arteries provides the abdominal parts of the esophagus with its arterial supply?

(A) cystic
(B) gastroduodenal
(C) left gastric
(D) hepatic
(E) left gastroepiploic

38. Rugae are located in which of the following structures?

(A) duodenum
(B) stomach
(C) cecum
(D) ileum
(E) transverse colon

39. The left gastro-omental artery arises from which of the following arteries?

(A) splenic
(B) hepatic
(C) gastroduodenal

(D) left gastric
(E) right gastric

40. The sympathetic nerve supply of the stomach arises from which of the following cord segments?

(A) T1–T5
(B) T6–T9
(C) T6–T12
(D) L1–L3
(E) T10–L2

41. The first part of the duodenum is located at which of the following vertebral levels?

(A) T10
(B) L2
(C) L1
(D) L5
(E) L3

42. The bile and pancreatic ducts enter which of the following structures?

(A) stomach
(B) 2nd portion of the duodenum
(C) cecum
(D) ileum
(E) liver

43. Which of the following structures crosses over the inferior or horizontal (third) portion of the duodenum?

(A) pancreas
(B) hepatic artery
(C) portal vein
(D) superior mesenteric artery

44. The duodenojejunal junction is supported by the attachment of which of the following structures?

(A) suspensory muscle of the duodenum (ligament of Treitz)
(B) falciform ligament
(C) hepatoduodenal ligament
(D) greater omentum
(E) transverse mesocolon

45. The superior anterior and posterior pancreaticoduodenal arteries arise from which of the following arteries?

(A) right colic
(B) ileocolic
(C) gastroduodenal
(D) hepatic
(E) splenic

46. The root of the mesentery crosses all of the following structures EXCEPT

(A) ascending and horizontal parts of the duodenum
(B) abdominal aorta
(C) inferior vena cava
(D) right ureter
(E) splenic artery

47. The superior mesenteric and splenic veins unite to form the portal vein posterior to which of the following structures?

(A) right kidney
(B) neck of the pancreas
(C) pylorus of stomach
(D) 2nd portion of the duodenum
(E) spleen

48. The sympathetic fibers in the nerves to the jejunum and ileum originate in which of the following spinal cord segments?

(A) C5–T1
(B) T1–T5
(C) T5–T9
(D) T9–T12
(E) L1–L2

49. Circular folds (plicae circulares) are characteristic of which of the following structures?

(A) transverse colon
(B) stomach
(C) jejunum
(D) cecum
(E) sigmoid colon

50. Omental appendices are located on which of the following structures?

(A) stomach
(B) duodenum
(C) ileum
(D) ascending colon
(E) liver

51. The are no teniae coli in which of the following structures?

(A) ascending colon
(B) transverse colon
(C) descending colon
(D) sigmoid colon
(E) appendix

52. The appendicular artery is a branch of which of the following arteries?

(A) inferior mesenteric
(B) inferior epigastric
(C) ileocolic
(D) testicular
(E) renal

53. Which of the following structures can be located deep to a point that is one-third of the way along the oblique line, joining the right anterior superior iliac spine to the umbilicus?

(A) gallbladder
(B) spleen
(C) right kidney
(D) appendix
(E) urinary bladder

54. Which of the following structures receives parasympathetic innervation from the pelvic splanchnic nerves?

(A) appendix
(B) sigmoid colon
(C) ileum
(D) ascending colon
(E) duodenum

55. The rectum is continuous with the sigmoid colon at the level of which of the following vertebrae?

(A) L3
(B) L5
(C) S3
(D) S5
(E) coccygeal 1

56. All of the following statements concerning the spleen are correct EXCEPT

(A) largest of the lymphatic organs
(B) associated posteriorly with the left 9th through 11th ribs
(C) located retroperitoneally
(D) normally, does not descend inferior to the costal region
(E) varies considerably in size, weight, and shape

57. The splenic artery usually follows a tortuous course along which of the following structures?

(A) left kidney
(B) greater curvature of the stomach
(C) pancreas
(D) transverse colon
(E) cecum

58. The head of the pancreas is embraced by which of the following structures?

(A) stomach
(B) spleen
(C) cecum
(D) C-shaped curve of the duodenum
(E) transverse mesocolon

59. The head of the pancreas rests posteriorly on which of the following structures?

(A) left renal vein
(B) superior vena cava
(C) splenic artery
(D) duodenum
(E) superior mesenteric artery

60. The main pancreatic duct and the bile duct unite to form which of the following structures?

(A) common bile duct
(B) hepatic duct

(C) accessory pancreatic duct
(D) cystic duct
(E) hepatopancreatic ampulla

61. The round ligament of the liver is the fibrous remnant of which of the following structures?
(A) umbilical vein
(B) ductus venosus
(C) ductus arteriosus
(D) umbilical artery
(E) urachus

62. The porta hepatis gives passage to all of the following structures EXCEPT
(A) portal vein
(B) hepatic artery
(C) hepatic ducts
(D) lymphatic vessels
(E) cystic artery

63. Which of the following ligaments encloses the portal triad?
(A) hepatoduodenal
(B) hepatogastric
(C) gastrocolic
(D) hepatorenal
(E) gastrosplenic

64. The portion of the hepatic artery extending between the celiac trunk and the gastroduodenal artery is known as the
(A) proper hepatic
(B) common hepatic
(C) right hepatic
(D) left hepatic
(E) middle hepatic

65. The hepatic veins drain into which of the following structures?
(A) liver
(B) inferior vena cava
(C) spleen
(D) portal vein
(E) superior vena cava

66. The spiral valve is located in which of the following structures?
(A) head of pancreas
(B) pylorus
(C) neck of gallbladder
(D) cecum
(E) rectum

67. The cystic artery commonly arises from the right hepatic artery in the angle between the common hepatic duct and which of the following structures?
(A) cystic duct
(B) celiac trunk
(C) portal vein
(D) proper hepatic artery
(E) gastroduodenal artery

68. Which of the following veins, when dilated, produces caput medusae?
(A) proper hepatic
(B) splenic
(C) cystic
(D) paraumbilical
(E) rectal

69. Inferiorly, the posterior surfaces of the kidney are related to all of the following structures EXCEPT
(A) subcostal nerve
(B) iliohypogastric nerve
(C) ilioinguinal nerve
(D) 2nd portion of the duodenum
(E) quadratus lumborum

70. The renal papillae empty into which of the following structures?
(A) renal vein
(B) ureter
(C) minor calyces
(D) renal pyramid
(E) renal column

71. All of the following statements concerning the renal hilum are correct EXCEPT

(A) the renal vein is anterior to the renal artery
(B) the renal artery is anterior to the renal pelvis
(C) it is the entrance to the renal sinus
(D) it is located on the concave medial margin of the kidney
(E) it contains the renal pyramids

72. The suprarenal glands are located between the superomedial aspects of the kidneys and which of the following structures?

(A) neck of the pancreas
(B) diaphragm
(C) quadrate lobe of the liver
(D) mesentery
(E) 1st part of the duodenum

73. Which of the following structures is related to the spleen, stomach, pancreas, and the left crus of the diaphragm?

(A) left suprarenal gland
(B) left kidney
(C) left gonadal vein
(D) transverse mesocolon
(E) abdominal aorta

74. All of the following statements concerning the suprarenal cortex are correct EXCEPT

(A) derives from mesoderm
(B) secretes corticosteroids
(C) secretes androgens
(D) associated with the sympathetic nervous system
(E) causes the kidney to retain sodium

75. The superior suprarenal arteries are branches of which of the following arteries?

(A) abdominal aorta
(B) renal
(C) inferior phrenic
(D) celiac trunk
(E) superior mesenteric

76. Which of the following muscles is considered to be the chief muscle of inspiration?

(A) internal intercostal
(B) external intercostal
(C) diaphragm
(D) scalene
(E) sternocleidomastoid

77. All of the following statements concerning the central tendon of the diaphragm are correct EXCEPT

(A) It has no bony attachments.
(B) It is incompletely divided into three leaves.
(C) It is perforated by the aorta.
(D) It lies near the center of the diaphragm.
(E) It is perforated by the inferior vena cava.

78. The crura of the diaphragm are musculotendinous bundles that arise from which of the following structures?

(A) posterior longitudinal ligament
(B) sternum
(C) bodies of lumbar vertebrae L1, L2, and L3
(D) renal fascia
(E) psoas major muscle

79. The nerves of the kidneys and suprarenal glands are derived from which of the following plexuses?

(A) celiac
(B) lumbar
(C) inferior mesenteric
(D) sacral
(E) inferior hypogastric

80. The lateral arcuate ligaments are formed from thickenings from which of the following muscular fasciae?

(A) psoas major
(B) quadratus lumborum
(C) transversus abdominis
(D) rectus abdominis
(E) sternalis

81. All of the following structures may herniate into the thoracic cavity when there is a traumatic diaphragmatic hernia EXCEPT

(A) stomach
(B) kidney
(C) intestine
(D) mesentery
(E) spleen

82. All of the following structures pass through the esophageal hiatus EXCEPT

(A) posterior vagal trunk
(B) esophageal branches of the left gastric vessels
(C) anterior vagal trunk
(D) thoracic duct
(E) esophagus

83. The greater and lesser splanchnic nerves pass through the diaphragm via which of the following structures?

(A) sternocostal foramen
(B) aortic hiatus
(C) diaphragmatic crus
(D) vena caval foramen
(E) medial arcuate ligament

84. The parasympathetic root of the celiac plexus is a branch of which of the following?

(A) greater splanchnic
(B) pelvic splanchnic
(C) lumbar splanchnic
(D) posterior vagal trunk
(E) lesser splanchnic

85. The bifurcation of the abdominal aorta occurs at the level of which of the following structures?

(A) crest of the ilium
(B) inguinal ligament
(C) pubic tubercle
(D) symphysis pubis
(E) obturator foramen

86. The inferior vena cava begins anterior to which of the following structures?

(A) right crus of the diaphragm
(B) right kidney
(C) 5th lumbar vertebra
(D) crest of the ilium
(E) bifurcation of the aorta

87. The cisterna chyli is the inferior end of which of the following structures?

(A) inferior vena cava
(B) abdominal aorta
(C) renal vein
(D) testicular artery
(E) thoracic duct

88. How are the pelvic splanchnic nerves distinct from other splanchnic nerves?

(A) derived from ventral primary rami of L2, L3, and L4
(B) derived from the sympathetic trunks
(C) convey preganglionic parasympathetic fibers to the inferior hypogastric plexus
(D) provide postganglionic sympathetic innervation to the cecum
(E) convey postganglionic parasympathetic innervation to the ascending colon

89. All of the following statements concerning the psoas major muscle are correct EXCEPT

(A) It passes inferolaterally, deep to the inguinal ligament.
(B) It inserts on the lesser trochanter of the femur.
(C) The pelvic splanchnic nerves are embedded in the posterior part of the psoas.
(D) It is a long, thick and fusiform muscle.
(E) Its name stems from a Greek word meaning "muscle of the loin."

90. Which of the following statements correctly applies to the suprarenal medulla?

(A) derived from mesoderm
(B) secretes corticosteroids and androgens
(C) contains chromaffin cells
(D) secretes acetylcholine
(E) produces hormones that cause the kidney to retain sodium and water in response to stress

91. The anatomical left and right lobes of the liver are separated on the diaphragmatic surface of the liver by which of the following structures?

(A) fissure for the round ligament of the liver
(B) fissure for the ligamentum venosum
(C) falciform ligament
(D) porta hepatis
(E) lesser omentum

92. All of the following statements concerning the spleen are correct EXCEPT

(A) It is the largest branch of the celiac trunk.
(B) It follows a tortuous course along the inferior border of the pancreas.
(C) It divides into five or more branches that enter the hilum of the spleen.
(D) It runs anterior to the left kidney.
(E) It follows posterior to the omental bursa.

93. In the developing embryo, the midgut rotates 270 degrees around which of the following structures?

(A) superior mesenteric artery
(B) celiac trunk
(C) splenic artery
(D) left renal artery
(E) inferior vena cava

94. Which of the following structures is avascular?

(A) superior ileocecal fold
(B) inferior ileocecal fold
(C) mesoappendix
(D) appendix
(E) cecum

95. Which of the following statements correctly relates to the 3rd part of the duodenum?

(A) It is supported by the suspensory ligament of the duodenum.
(B) The bile and pancreatic ducts enter its posteromedial wall.
(C) It is crossed by the superior mesenteric artery and vein.
(D) It attaches to the hepatoduodenal ligament.
(E) It lies to the right and runs parallel to the inferior vena cava.

96. The left gastro-omental artery arises from which of the following arteries?

(A) celiac trunk
(B) right gastric
(C) gastroduodenal
(D) splenic
(E) hepatic

97. Rugae are located in which of the following structures?

(A) duodenum
(B) jejunum
(C) ileum
(D) cecum
(E) stomach

98. All of the following statements concerning the esophagus are correct EXCEPT

(A) It extends from the pharynx to the stomach.
(B) It is crossed by the arch of the aorta.
(C) It is crossed by the right main bronchus.
(D) It passes through the esophageal hiatus.
(E) It normally has four constrictions.

99. Digestion occurs mainly in which of the following structures?

(A) cecum and ascending colon
(B) transverse colon
(C) stomach and duodenum
(D) jejunum and ileum
(E) transverse and sigmoid colon

100. Most reabsorption of water occurs in which of the following structures?

(A) stomach
(B) jejunum
(C) sigmoid colon
(D) ascending colon
(E) rectum

DIRECTIONS (Questions 101 through 110): Identify the anatomical features indicated on the art below.

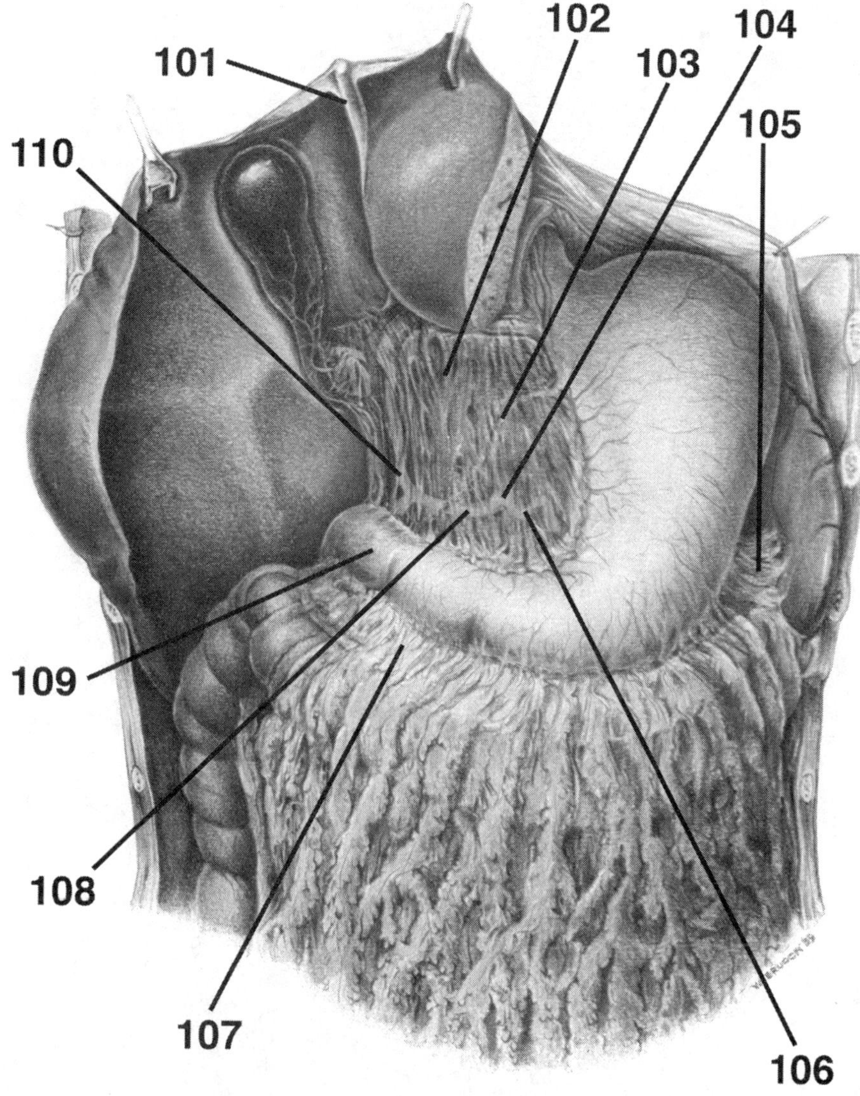

DIRECTIONS (Questions 111 through 115): Identify the anatomical features indicated on the art below.

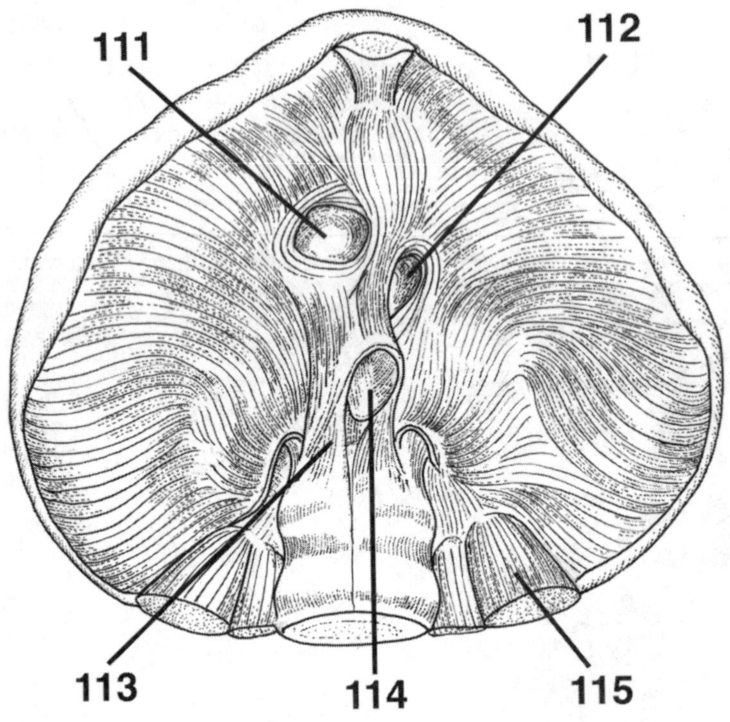

DIRECTIONS (Questions 116 through 120): Identify the anatomical features indicated on the art below.

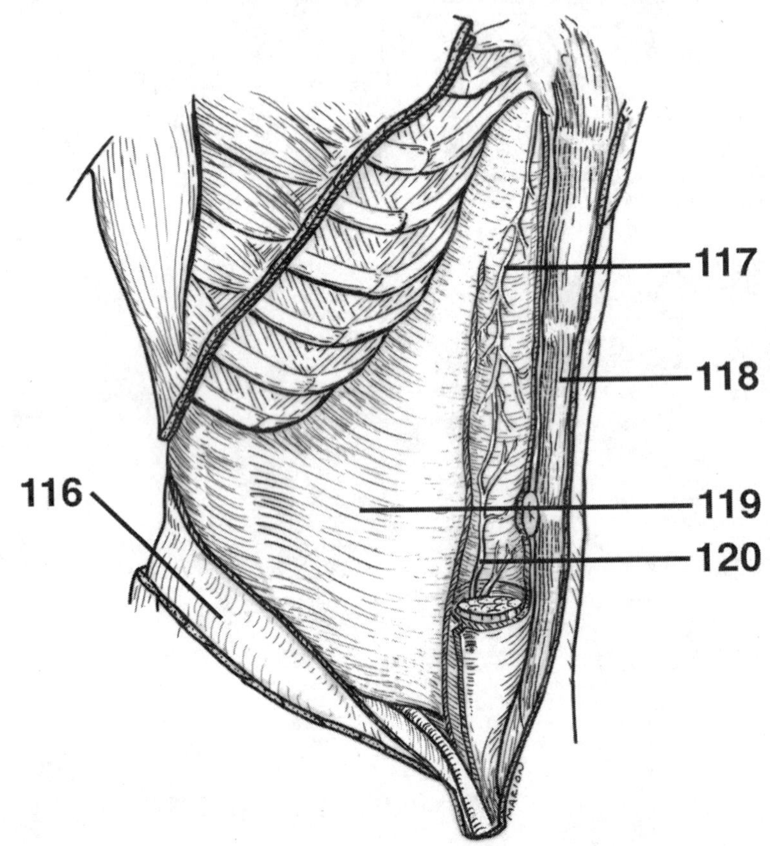

DIRECTIONS (Questions 121 through 125): Identify the anatomical features indicated on the art below.

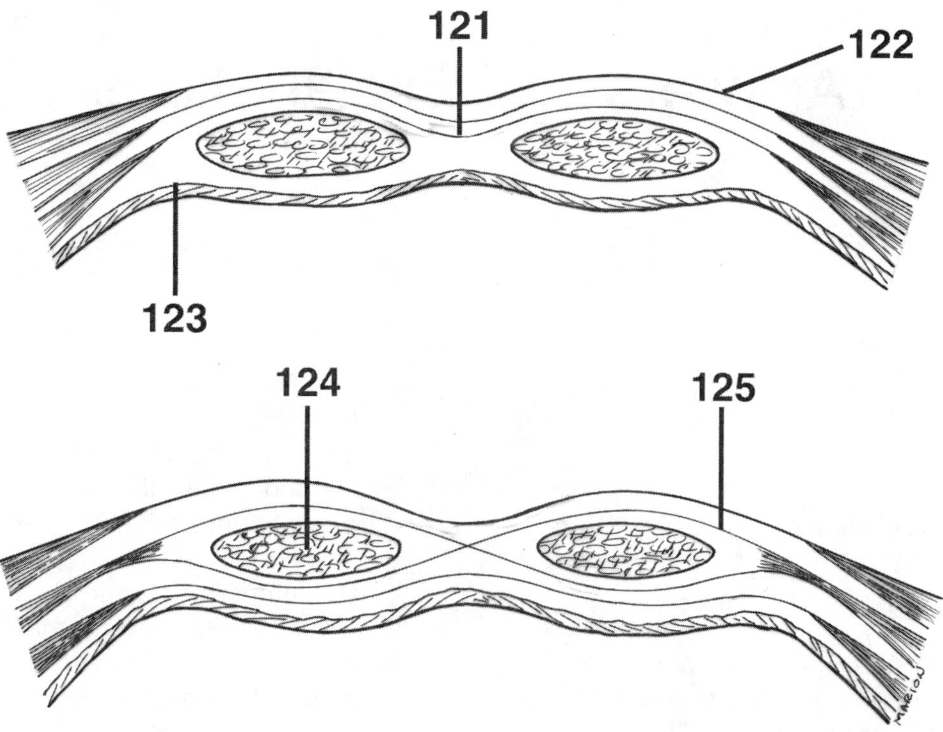

DIRECTIONS (Questions 126 through 130): Identify the anatomical features indicated on the art below.

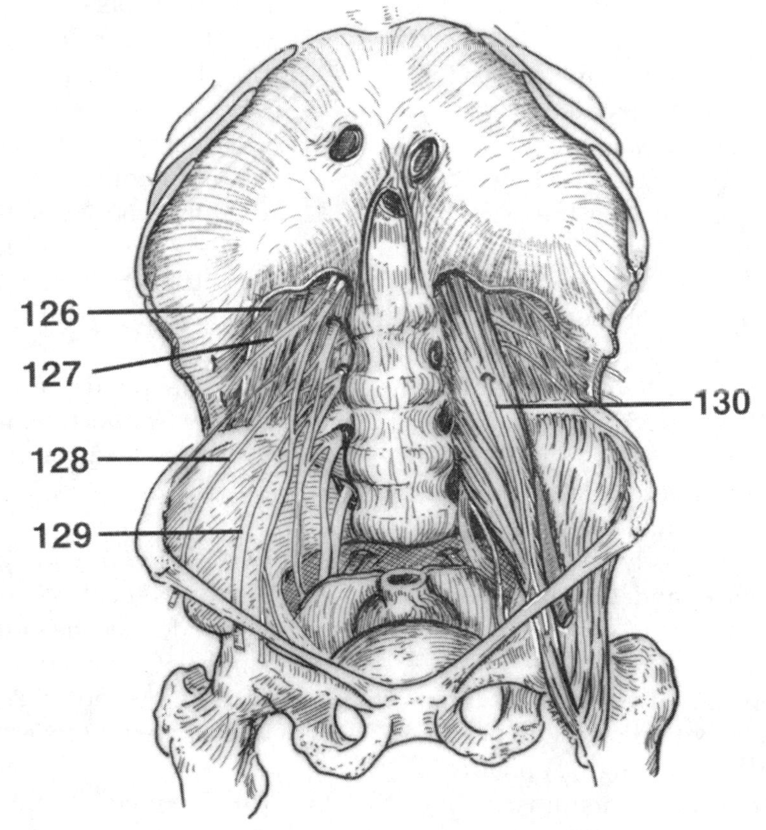

Answers and Explanations

1. **(D)** The transumbilical and median planes divide the abdomen into four quadrants *(Moore, p 176).*

2. **(D)** The midclavicular planes pass through the midpoint of the clavicles to the midinguinal points *(Moore, p 176).*

3. **(C)** Transversalis fascia lines most of the abdominal wall and covers the deep surface of the transverse abdominal muscle. The deep fascia invests the external abdominal oblique muscle. Both Camper's and Scarpa's fascia are located in the inferior part of the abdominal subcutaneous tissue *(Moore, pp 178–179).*

4. **(E)** The parietal peritoneum is internal to the transversalis fascia and is separated from it by a variable amount of endoabdominal (extraperitoneal) fat. Camper's fascia and the deep fasciae are all superficial to the extraperitoneal fat *(Moore, p 179).*

5. **(B)** The superficial fibers of the external abdominal oblique arising from the middle and lower ribs interdigitate with only those of the serratus anterior *(Moore, p 180).*

6. **(B)** The anterolateral abdominal wall includes the cartilages of the 7th–10th ribs and xiphoid process superiorly. Inferiorly it is bounded by the inguinal ligament and the pelvic bone *(Moore, p 178).*

7. **(C)** The aponeuroses of these muscles interlace at the linea alba with their fellows of the opposite side to form the tough, aponeurotic tendinous sheath of the rectus muscle, and the rectus sheath *(Moore, p 180).*

8. **(D)** The contents of the rectus sheath include the pyramidalis and rectus abdominis muscles, the superior and inferior epigastric arteries and veins, the lymphatics, and ventral primary rami of T7–T12 nerves *(Moore, p 180).*

9. **(B)** Inferiorly, the inferior margin of the external oblique aponeurosis thickens and folds back on itself to form the inguinal ligament, a fibrous band extending between the anterior superior iliac spine and the pubic tubercle *(Moore, p 180).*

10. **(A)** Some fibers of the inguinal ligament pass upward to cross the linea alba and blend with the lower fibers of the contralateral aponeurosis. These fibers form the reflected inguinal ligament *(Moore, p 180).*

11. **(C)** The neurovascular plane of the anterolateral abdominal wall contains the thoracoabdominal nerves, cutaneous branches T7–T11, ventral ramus of T12, iliohypogastric and ilioinguinal nerves, lumbar arteries, and the deep circumflex iliac artery *(Moore, p 182).*

12. **(D)** The rectus muscle is anchored transversely by attachments to the anterior layer of the rectus sheath and three or more tendinous intersections *(Moore, p 183).*

13. **(A)** The inferior limit of the posterior lamina of the rectus sheath is marked by the arcuate line, which defines the point at which the posterior lamina of the internal oblique and the aponeurosis of the transverse abdominal become part of the anterior rectus sheath *(Moore, p 184).*

14. **(B)** The median umbilical fold represents the remnant of the urachus; the medial umbilical

folds are remnants of the occluded fetal umbilical arteries; and the lateral umbilical folds cover the inferior epigastric vessels *(Moore, p 191).*

15. **(B)** The medial inguinal fossae between the medial and lateral umbilical folds, the inguinal triangles, are potential sites for the less common direct inguinal hernias. The lateral inguinal fossae, lateral to the lateral umbilical folds, are potential sites for the most common type of hernia in the lower abdominal wall—indirect inguinal hernia *(Moore, p 191).*

16. **(B)** The inguinal canal contains the spermatic cord in the male and the round ligament in the female. The inguinal canal also contains blood and lymphatic vessels and the ilioinguinal nerve in both sexes *(Moore, p 193).*

17. **(E)** The deep inguinal ring (entrance to the inguinal canal) is the site of an outpouching of the transversalis fascia approximately 1.2 cm superior to the middle of the inguinal ligament *(Moore, p 193).*

18. **(D)** The inguinal ligament is reinforced in its most medial part by the lacunar ligament, a reflected part or extension from the deep aspect of the inguinal ligament to the pectineal line of the pecten pubis *(Moore, p 193).*

19. **(B)** The iliopubic tract is the thickened inferior margin of the transversalis fascia that appears as a fibrous band running parallel and deep to the inguinal ligament. The iliopubic tract demarcates the inferior edge of the deep inguinal ring *(Moore, p 193).*

20. **(C)** The testes develop in the extraperitoneal connective tissue in the superior lumbar region of the posterior abdominal wall *(Moore, p 193).*

21. **(D)** The gubernaculum is a fibrous cord connecting the primordial testis to the anterolateral wall at the site of the future deep inguinal ring. The gubernaculum is represented postnatally by the scrotal ligament, which extends from the testis to the skin of the scrotum *(Moore, p 194).*

22. **(C)** The cremasteric fascia contains loops of cremaster muscle, which is formed by the lowermost fascicles of the internal oblique muscle arising from the inguinal ligament. The cremasteric fascia is derived from the deep and superficial fascia of the internal oblique *(Moore, p 198).*

23. **(A)** The cremaster muscle is innervated by the genital branch of the genitofemoral nerve *(Moore, p 198).*

24. **(C)** The artery of the ductus deferens arises from the inferior vesicle *(Moore, p 198).*

25. **(A)** Genital branch of the genitofemoral, anterior scrotal branches of the ilioinguinal, posterior scrotal branches of the pudendal, and perineal branches of the posterior femoral cutaneous provide the innervation to the scrotum *(Moore, p 201).*

26. **(C)** The epididymis lies on the posterior surface of the testis, which is covered by the tunica vaginalis except at its posterior margin *(Moore, p 201).*

27. **(B)** The testes are covered with a tough fibrous coat know as the tunica albuginea *(Moore, p 201).*

28. **(D)** The tunica vaginalis is a closed peritoneal sac partially surrounding the testis, which represents the closed-off distal part of the embryonic processus vaginalis *(Moore, p 202).*

29. **(D)** The pampiniform plexus is part of the thermoregulatory system of the testis, helping to keep this gland at a constant temperature *(Moore, p 202).*

30. **(B)** The autonomic nerves of the testis arise as the testicular plexus of nerves on the testicular artery, which contains vagal parasympathetic fibers and sympathetic fibers from the T7 segment of the spinal cord *(Moore, p 202).*

31. **(E)** The peritoneal cavity is a potential space of capillary thinness between the parietal and visceral layers of peritoneum. It contains no organs but rather a thin film of peritoneal fluid that lubricates the peritoneal surfaces *(Moore, p 210).*

32. **(D)** The lesser omentum connects the lesser curvature of the stomach and the proximal part of the duodenum to the liver *(Moore, p 213).*

33. **(C)** The greater omentum has considerable mobility and moves around the peritoneal cavity with peristaltic movements of the viscera. It wraps itself around an inflamed organ such as the appendix, walling it off and thereby protecting other viscera from it. For this reason, the great omentum is referred to as the "policeman of the peritoneal cavity."

34. **(D)** The liver is connected to the duodenum by the hepatoduodenal ligament (the thickened free edge of the lesser omentum that conducts the portal triad: portal vein, hepatic artery, and bile duct) *(Moore, p 213).*

35. **(C)** The caudate lobe of the liver, covered with visceral peritoneum, forms the superior boundary of the omental foramen (epiploic foramen, or foramen of Winslow), an opening situated posterior to the free edge of the lesser omentum (hepatoduodenal ligament).

36. **(D)** The esophagus has internal circular and external longitudinal layers of muscle. In its superior third, the external layer consists of skeletal muscle, the inferior third is composed of smooth muscle, and the middle third is made up of both types of muscle *(Moore, p 224).*

37. **(C)** The arterial supply of the abdominal part of the esophagus is from the left gastric artery, a branch of the celiac trunk, and the left inferior phrenic artery *(Moore, p 225).*

38. **(B)** When the gastric mucosa is contracted, it is thrown into longitudinal ridges known as gastric folds or rugae *(Moore, p 227).*

39. **(A)** The left gastro-omental artery arises from the splenic artery and courses along the greater curvature to anastomose with the right gastro-omental artery *(Moore, p 229).*

40. **(B)** The sympathetic nerve supply of the stomach from T6 through T9 segments of the spinal cord passes to the celiac plexus through the greater splanchnic nerve *(Moore, p 231).*

41. **(C)** The superior (1st) part of the duodenum lies anterolateral to the body of L1 vertebra *(Moore, p 237).*

42. **(B)** The bile and pancreatic ducts enter the posteromedial wall of the 2nd portion of the duodenum *(Moore, p 237).*

43. **(D)** The inferior or horizontal (third) portion of the duodenum is crossed by the superior mesenteric artery and vein and the root of the mesentery of the jejunum and ileum *(Moore, p 237).*

44. **(A)** The duodenojejunal junction is supported by the suspensory muscle of the duodenum (ligament of Treitz) *(Moore, p 237).*

45. **(C)** The gastroduodenal artery and its branches, the superior anterior and posterior pancreaticoduodenal arteries, supply the duodenum proximal to the entry of the bile duct into the descending (second) portion of the duodenum *(Moore, p 241).*

46. **(E)** The root of the mesentery crosses (successively) the ascending and horizontal parts of the duodenum, abdominal aorta, inferior vena cava, right ureter, right psoas major, and right testicular or ovarian vessels *(Moore, p 244).*

47. **(B)** The superior mesenteric and splenic veins unite to form the portal vein posterior to the neck of the pancreas *(Moore, p 244).*

48. **(C)** The sympathetic fibers in the nerves to the jejunum and ileum originate in the T5 through T9 segments of the spinal cord *(Moore, p 244).*

49. **(C)** The jejunum contains circular folds along with the duodenum. The circular folds are absent in the ileum and large intestine *(Moore, p 244).*

50. **(D)** The large intestine can be distinguished from the small intestine by teniae coli, haustra, and omental appendices *(Moore, p 249).*

51. **(E)** There are no teniae in the appendix or rectum *(Moore, p 249).*

52. **(C)** The appendix is supplied by the appendicular artery, a branch of the ileocolic artery *(Moore, p 251).*

53. **(D)** The base of the appendix lies deep to a point that is one-third of the way along the oblique line, joining the right anterior superior iliac spine to the umbilicus (spinoumbilical or McBurney's point) *(Moore, p 251).*

54. **(B)** The pelvic splanchnic nerves provide the parasympathetic innervation for the distal one-third of the transverse colon, descending colon, and rectum *(Moore, p 255).*

55. **(C)** The rectum, the fixed terminal part of the large intestine, is continuous with the sigmoid colon at the level of S3 vertebra *(Moore, p 255).*

56. **(C)** The spleen is the largest of the lymphatic organs and is a mobile structure entirely surrounded by peritoneum except at the hilum. It lies beneath the 9th, 10th, and 11th ribs and does not normally descend inferior to the costal margin. The spleen varies considerably in size, weight, and shape *(Moore, p 256).*

57. **(C)** The splenic artery, the largest branch of the celiac trunk, follows a tortuous course posterior to the omental bursa, anterior to the left kidney, and along the superior border of the pancreas *(Moore, p 257).*

58. **(D)** The head of the pancreas is embraced by the C-shaped curve of the duodenum to the right of the superior mesenteric vessels *(Moore, p 257).*

59. **(A)** The head of the pancreas rests posteriorly on the inferior vena cava, right renal artery and vein and left renal vein *(Moore, p 259).*

60. **(D)** The main pancreatic duct and the bile duct unite to form a short, dilated hepatopancreatic ampulla, which opens into the descending part of the duodenum at the summit of the major duodenal papilla *(Moore, p 261).*

61. **(A)** The round ligament of the liver is the fibrous remnant of the umbilical vein that carried well-oxygenated and nutrient-rich blood from the placenta to the fetus *(Moore, p 265).*

62. **(E)** The porta hepatis gives passage to the portal vein, hepatic artery, hepatic nerve plexus, hepatic ducts, and lymphatic vessels *(Moore, p 265).*

63. **(A)** The portion of the lesser omentum extending between the porta hepatis of the liver and the duodenum (the hepatoduodenal ligament) encloses the portal triad *(Moore, p 265).*

64. **(B)** The common hepatic artery extends from the celiac trunk to the origin of the gastroduodenal artery. The proper hepatic artery extends from the origin of the gastroduodenal artery to its bifurcation into right and left hepatic branches *(Moore, p 265).*

65. **(B)** The hepatic veins, formed by the union of the central veins of the liver, open into the inferior vena cava just inferior to the diaphragm *(Moore, p 266).*

66. **(C)** The mucosa of the neck of the gallbladder spirals into a fold known as the spiral valve, which keeps the cystic duct open so that bile can easily divert into the gallbladder when the distal end of the bile duct is closed, or so bile can pass to the duodenum as the gallbladder contracts *(Moore, p 274).*

67. **(A)** The cystic artery commonly arises from the right hepatic artery in the angle between the common hepatic duct and the cystic duct *(Moore, p 275).*

68. **(D)** The paraumbilical veins of the anterior abdominal wall (portal system) anastomosing with superficial epigastric veins (systemic system), when dilated, produce caput medusae *(Moore, p 277).*

69. **(D)** Inferiorly, the posterior surfaces of the kidney are related to the quadratus lumborum muscle and the subcostal, iliohypogastric, and ilioinguinal nerves and vessels *(Moore, p 280).*

70. **(C)** The renal papillae empty into the minor calices, which empty into major calices which in turn empty into the pelvis of the ureter *(Moore, p 280).*

71. **(E)** At the concave medial margin of each kidney is a vertical cleft, the renal hilum, where the renal artery enters and the renal vein and renal pelvis leave the renal sinus. At the hilum, the renal vein is anterior to the renal artery, which is anterior to the renal pelvis *(Moore, p 280).*

72. **(B)** The suprarenal glands are located between the superomedial aspects of the kidneys and the diaphragm *(Moore, p 285).*

73. **(A)** The semilunar left suprarenal gland is related to the spleen, stomach, pancreas, and left crus of the diaphragm *(Moore, p 285).*

74. **(D)** The suprarenal cortex derives from mesoderm and secretes corticosteroids and androgens. These hormones cause the kidneys to retain sodium and water in response to stress *(Moore, p 285).*

75. **(C)** The superior suprarenal arteries are provided by the inferior phrenic. The middle suprarenal arteries are provided by the abdominal aorta and the inferior suprarenal arteries are provided by the renal artery *(Moore, p 287).*

76. **(C)** The diaphragm, the chief muscle of inspiration, descends during inspiration. Only its central part moves because its periphery, as the fixed origin of the muscle, attaches to the inferior margin of the thoracic cage and the superior lumbar vertebrae *(Moore, p 289).*

77. **(C)** The central tendon of the diaphragm has no bony attachments and is incompletely divided into three leaves, resembling a wide cloverleaf. Although it lies near the center of the diaphragm, the central tendon is closer to the anterior part of the thorax. The inferior vena cava perforates the central tendon. The aortic hiatus is formed by the right and left muscular crura *(Moore, p 292).*

78. **(C)** The crura of the diaphragm are musculotendinous bundles that arise from the anterior surfaces of the bodies of the superior three lumbar vertebrae, the anterior longitudinal ligament, and the intervertebral discs *(Moore, p 291).*

79. **(A)** The nerves of the kidneys and suprarenal glands are derived from the celiac plexus, the lesser and least thoracic splanchnic nerves, and the aorticorenal ganglion *(Moore, p 290).*

80. **(B)** The diaphragm is attached on each side to the medial and lateral arcuate ligaments, which are thickenings of the fascia covering the psoas major and quadratus lumborum muscles, respectively *(Moore, p 292).*

81. **(B)** The stomach, spleen, mesentery, and intestine may herniate into the thorax when there is a traumatic diaphragmatic hernia *(Moore, p 295).*

82. **(D)** The esophageal hiatus is an oval aperture for the esophagus in the muscle of the right crus of the diaphragm at the level of the T10 vertebra. The esophageal hiatus also transmits the anterior and posterior vagal trunks, esophageal branches of the left gastric vessels, and a few lymphatics *(Moore, p 294).*

83. **(C)** The sympathetic trunks pass deep to the medial arcuate ligament. There are two small apertures in each crus of the diaphragm; one transmits the greater and the other the lesser splanchnic nerve *(Moore, p 295).*

84. **(D)** The parasympathetic root of the celiac plexus is a branch of the posterior vagal trunk that contains fibers from the right and left vagal nerves. The sympathetic roots of the plexus are the greater and lesser splanchnic nerves *(Moore, p 3020).*

85. **(A)** The bifurcation of the aorta can be located on the surface anatomy approximately 2.5 cm superior to the transpyloric plane to a point slightly inferior to and to the left of the umbilicus. Bifurcation is also indicated just to the left of the midpoint of a line joining the highest points of the iliac crests *(Moore, p 305).*

86. **(C)** The inferior vena cava begins anterior to the L5 vertebra by the union of the common iliac veins. The union occurs approximately 2.5 cm to the right of the median plane, inferior to the bifurcation of the aorta and posterior to the proximal part of the right common iliac artery *(Moore, p 307).*

87. **(E)** In a small proportion of individuals, the inferior end of the thoracic duct—which begins with the convergence of the main lymphatic ducts of the abdomen—takes the form of the commonly depicted, thin-walled sac or dilation known as the cisterna chyli *(Moore, p 308).*

88. **(C)** The pelvic splanchnic nerves are distinct from other splanchnic nerves in that they have nothing to do with the sympathetic trunks and are derived directly from ventral rami of spinal nerves S2 through S4. They also convey presynaptic parasympathetic fibers to the inferior hypogastric plexus *(Moore, p 302).*

89. **(C)** The psoas muscle is a long, thick, fusiform muscle that lies lateral to the lumbar vertebrae. Psoas is a Greek word meaning "muscle of the loin." The muscle passes inferolaterally, deep to the inguinal ligament to reach the lesser trochanter of the femur. The lumbar plexus of nerves is embedded in the posterior part of the psoas *(Moore, p 299).*

90. **(C)** The suprarenal medulla is derived from neural crest cells associated with the sympathetic nervous system. The chromaffin cells of the medulla are related to sympathetic (postganglionic) neurons in both derivation (neural crest cells) and function. These cells secrete catecholamines (mostly epinephrine) into the bloodstream in response to signals from presynaptic neurons *(Moore, p 286).*

91. **(C)** In the current terminology, the left liver includes the caudate lobe and most of the quadrate lobe. The anatomical left lobe is separated from these lobes on the visceral surface by the fissure for the round ligament of the liver and the fissure for the ligamentum venosum; on the diaphragmatic surface the anatomical left lobe is separated by the attachment of the falciform ligament *(Moore, p 265).*

92. **(B)** The splenic artery is the largest branch of the celiac trunk. It follows a tortuous course posterior to the omental bursa, anterior to the left kidney, and along the superior border of the pancreas. It divides into five or more branches that enter the hilum of the spleen *(Moore, p 257).*

93. **(A)** For several weeks the rapidly growing midgut, supplied by the superior mesenteric artery, is physiologically herniated into the proximal part of the umbilical cord. It is attached to the yolk sac by the yolk stalk. As it returns to the abdominal cavity, the midgut rotates 270 degrees around the axis of the superior mesenteric artery *(Moore, p 246).*

94. **(B)** The appendicular artery is located in the mesoappendix and the anterior cecal artery is located in the superior ileocecal fold. Both the appendix and cecum are vascular. The inferior ileocecal fold is avascular *(Moore, p 238).*

95. **(C)** The inferior or horizontal (3rd) part of the duodenum runs transversally to the left, passing over the inferior vena cava, aorta, and L3 vertebra. It is crossed by the superior mesenteric artery and vein and the root of the mesentery *(Moore, p 237).*

96. **(D)** The left gastro-omental artery arises from the splenic artery and courses along the greater curvature to anastomose with the right gastro-omental artery *(Moore, p 229).*

97. **(E)** The smooth surface of the gastric mucosa, mucous layer of the stomach, is thrown into longitudinal ridges when contracted. These are the gastric folds or rugae and most obvious along the greater curvature of the stomach *(Moore, p 227).*

98. **(C)** The esophagus is a muscular tube (approximately 25 cm long) that extends from the pharynx to the stomach. It normally has four constrictions and is crossed by the arch of the aorta and the left main bronchus *(Moore, p 221).*

99. **(C)** Digestion occurs mainly in the stomach and duodenum. Peristalsis—ring-like, contraction waves that begin around the middle of the stomach and move slowly toward the pylorus—is responsible for mixing the masticated food mass with gastric juices and for emptying the contents of the stomach into the duodenum *(Moore, p 218).*

100. **(D)** Most reabsorption occurs in the ascending colon *(Moore, p 218).*

101. falciform ligament

102. lesser omentum

103. left gastric artery

104. celiac trunk

105. gastrolienal ligament

106. splenic artery

107. gastrocolic ligament

108. common hepatic artery

109. pylorus

110. proper hepatic artery

111. caval opening

112. esophageal hiatus

113. right crus of diaphragm

114. aortic hiatus

115. quadratus lumborum

116. internal abdominal oblique

117. superior epigastric artery

118. rectus abdominis

119. transversus abdominis

120. inferior epigastric artery

121. transversus abdominis aponeurosis

122. external abdominal oblique aponeurosis

123. transversalis fascia

124. rectus abdominis

125. internal abdominal oblique aponeurosis

126. subcostal nerve

127. iliohypogastric nerve

128. lateral femoral cutaneous nerve

129. femoral nerve

130. genitofemoral nerve

CHAPTER 5

The Pelvis and Perineum

Questions

DIRECTIONS (Questions 1 through 100): Each of the numbered items or incomplete statements in this section is followed by answers or by completions of the statement. Select the ONE lettered answer or completion that is BEST in each case.

1. All of the following statements concerning the pelvic cavity are correct EXCEPT

(A) It is bounded posteriorly by the coccyx.
(B) It is bounded anteriorly by the pubic symphysis.
(C) The pelvic inlet forms the superior boundary.
(D) The pelvic outlet forms the inferior boundary.
(E) The musculofascial pelvic diaphragm closes the pelvic inlet.

2. All of the following bones contribute to the formation of the pelvic cavity EXCEPT

(A) ischium
(B) pelvis
(C) pubis
(D) coccyx
(E) sacrum

3. Which of the following structures is located between the ischial spine and the ischial tuberosity?

(A) obturator foramen
(B) lesser sciatic notch
(C) acetabular notch
(D) pubic arch
(E) arcuate line

4. The lateral part of the superior ramus of the pubis forms which of the following structures?

(A) iliopubic eminence
(B) pubic tubercle
(C) pecten pubis
(D) anterior inferior iliac spine
(E) acetabulum

5. All of the following structures provide boundaries for the pelvic inlet EXCEPT

(A) inferior ramus of the pubis
(B) sacral promontory
(C) anterior border of the ala of the sacrum
(D) arcuate line of the ilium
(E) pecten pubis

6. When a person is in the anatomical position, which of the following structures lie in the same vertical plane?

(A) sacral promontory and pubic tubercles
(B) anterior superior iliac spines and the anterior aspect of the pubic symphysis
(C) posterior superior iliac spines and the posterior aspect of the ischial tuberosity
(D) ischial spines and the posterior border of the obturator foramen
(E) superior pubic rami and the greater sciatic notch

7. Weak areas of the pelvis include all of the following EXCEPT

(A) ischial tuberosities
(B) alae of the ilium
(C) pubic rami
(D) sacroiliac joint
(E) acetabula

8. The pelvic floor is formed by all of the following muscles EXCEPT

(A) pubococcygeus
(B) coccygeus
(C) piriformis
(D) puborectalis
(E) iliococcygeus

9. Which of the following muscles covers and pads the lateral pelvic wall?

(A) obturator internus
(B) piriformis
(C) pubococcygeus
(D) iliococcygeus
(E) ischiococcygeus

10. All of the following statements concerning the sciatic nerve are correct EXCEPT

(A) It is the largest and broadest nerve in the body.
(B) It is formed by the dorsal rami of L4 to S3.
(C) It passes through the greater sciatic foramen.
(D) It exits the pelvis inferior to the piriformis muscle.
(E) It is one of the two main nerves of the sacral plexus.

11. All of the following statements concerning the pudendal nerve are correct EXCEPT

(A) It is derived from the anterior divisions of the ventral rami of S2 through S4.
(B) It accompanies the internal pudendal artery.
(C) It leaves the pelvis through the greater sciatic foramen.
(D) It leaves the pelvis through the greater sciatic foramen inferior to the piriformis and coccygeus muscles.
(E) It is the chief sensory nerve of the external genitalia.

12. Which of the following nerves exits the pelvis through the greater sciatic foramen, superior to the piriformis?

(A) sciatic
(B) pudendal
(C) superior gluteal
(D) lumbosacral trunk
(E) obturator

13. Which of the following statements concerning the sacral sympathetic trunks is correct?

(A) Usually has four sympathetic ganglia.
(B) Descends on the pelvic surface of the ischium.
(C) Ends as the dorsal nerve of the penis or clitoris.
(D) Passes through the obturator canal.
(E) Pierces the sacrotuberous ligament.

14. The inferior hypogastric plexus receives fibers from which of the following nerves?

(A) pudendal and obturator
(B) sciatic and superior gluteal
(C) inferior gluteal and lumbosacral
(D) pelvic splanchnic and hypogastric
(E) sacral sympathetic and obturator

15. All of the following arteries enter the true pelvis EXCEPT

(A) internal iliac
(B) median sacral
(C) superior rectal
(D) ovarian
(E) testicular

16. Which of the following arteries is considered to be the artery of the pelvis?

(A) obturator
(B) pudendal

(C) uterine
(D) internal iliac
(E) sacral

17. Which of the following arteries arises from the anterior division of the internal iliac?

(A) superior rectal
(B) iliolumbar
(C) superior gluteal
(D) gonadal
(E) obturator

18. All of the following statements concerning the ureters are correct EXCEPT

(A) They are retroperitoneal.
(B) They cross the pelvic brim anterior to the bifurcation of the common iliac arteries.
(C) They are fibrous tubes connecting the kidneys to the urinary bladder.
(D) Their superior halves lie in the abdomen and the inferior halves lie in the pelvis.
(E) Their oblique passage through the bladder wall forms a one-way "flap valve."

19. The uvula of the bladder is located in which of the following locations?

(A) retropubic space
(B) trigone of the bladder
(C) apex of the bladder
(D) fundus of the bladder
(E) pubovesical ligament

20. Parasympathetic fibers to the bladder are derived from which of the following nerves?

(A) pelvic splanchnic
(B) greater splanchnic
(C) T11-L2
(D) superior hypogastric plexus
(E) sacral plexus

21. Which of the following parts of the male urethra is the widest and most dilatable?

(A) preprostatic
(B) prostatic
(C) membranous
(D) spongy
(E) external urethral meatus

22. Which of the following structures opens into the prostatic sinus?

(A) prostatic utricle
(B) ejaculatory ducts
(C) prostatic ductules
(D) seminal vesicles
(E) bulbourethral glands

23. The paraurethral glands open into which of the following structures?

(A) near the external urethral orifice
(B) in the neck of the bladder
(C) in the prostatic utricle
(D) in the seminal colliculus
(E) in the ejaculatory ducts

24. All of the following statements concerning the ductus deferens are correct EXCEPT

(A) It begins in the head of the epididymis.
(B) It ascends in the spermatic cord.
(C) It passes through the inguinal canal.
(D) It joins the duct of the seminal vesicle to form the ejaculatory duct.
(E) It descends medial to the ureter and seminal vesicle.

25. All of the following statements concerning the seminal vesicles are correct EXCEPT

(A) The peritoneum of the rectovesical pouch separates the superior ends of the seminal vesicles from the rectum.
(B) They do not store sperm.
(C) They are elongated structures that lie between the fundus of the bladder and the rectum.
(D) The inferior ends of the seminal vesicles are separated from the rectum by the rectovesical septum.
(E) Bulbourethral glands empty into the seminal vesicles.

26. The posterior fornix is the deepest part of which of the following structures?
   (A) urinary bladder
   (B) vagina
   (C) deep perineal space
   (D) urethra
   (E) uterus

27. All of the following statements concerning the uterus are correct EXCEPT
   (A) It is anteverted.
   (B) It is anteflexed.
   (C) The position of the uterus is fixed.
   (D) It is divisible into two main parts.
   (E) The body of the uterus lies between the layers of the broad ligament.

28. Which of the following parts of the uterus protrudes into the uppermost vagina?
   (A) round ligament
   (B) body
   (C) fundus
   (D) isthmus
   (E) cervix

29. The rounded vaginal part of the cervix extends into the vagina and communicates with it through which of the following structures?
   (A) uterine tubes
   (B) external os
   (C) ureter
   (D) urethra
   (E) internal os

30. Laterally, the peritoneum of the broad ligament is prolonged superiorly over the ovarian vessels as which of the following structures?
   (A) mesovarium
   (B) mesosalpinx
   (C) mesometrium
   (D) suspensory ligament of the ovary
   (E) transverse cervical (cardinal) ligaments

31. The oocytes expelled from the ovaries usually are fertilized in which of the following areas of the uterine tubes?
   (A) infundibulum
   (B) ampulla
   (C) isthmus
   (D) uterine
   (E) fimbria

32. The distal end of the ovary connects to the lateral wall of the pelvis by which of the following structures?
   (A) round ligament
   (B) ligament of the ovary
   (C) suspensory ligament of the ovary
   (D) transverse cervical ligament
   (E) lateral cervical ligament

33. Which of the following nerves provides parasympathetic innervation to the ovaries?
   (A) pudendal
   (B) pelvic splanchnic
   (C) vagus
   (D) obturator
   (E) superior hypogastric

34. The rectosigmoid junction lies anterior to which of the following structures?
   (A) S3 vertebra
   (B) prostate
   (C) obturator foramen
   (D) bladder
   (E) pararectal fossae

35. All of the following osseofibrous structures mark the boundaries of the perineum EXCEPT
   (A) pubic symphysis
   (B) inferior pubic rami
   (C) sacrospinous ligament
   (D) ischial tuberosities
   (E) ischial rami

36. The perineum is divided into two triangles by drawing a transverse line between which of the following structures?

(A) anterior ends of the ischial tuberosities
(B) coccyx to pubic tubercles
(C) inferior iliac spines to pubic symphysis
(D) medial ends of inguinal ligament to tip of coccyx
(E) sacrum to pubic symphysis

37. The perineal body is the site of convergence of all of the following muscles EXCEPT

(A) ischiocavernous
(B) bulbospongiosus
(C) superficial transverse perineal
(D) deep transverse perineal
(E) external anal sphincter

38. In males, the superficial perineal pouch contains which of the following structures?

(A) prostate
(B) seminal vesicles
(C) membranous urethra
(D) neck of the bladder
(E) ischiocavernous muscle

39. In females, the deep perineal pouch contains which of the following structures?

(A) clitoris
(B) greater vestibular glands
(C) bulbourethral glands
(D) bulbs of the vestibule
(E) external urethral sphincter

40. The pudendal canal is a space within which of the following structures?

(A) the deep perineal space
(B) the superficial perineal space
(C) the obturator fascia
(D) the broad ligament
(E) the urogenital triangle

41. The pudendal nerve innervates which of the following structures?

(A) ovaries
(B) testes
(C) uterus
(D) levator ani
(E) muscles of the deep and superficial perineal pouch

42. The anal columns contain which of the following structures?

(A) internal pudendal vessels
(B) pudendal nerve
(C) superior rectal vessels
(D) obturator nerve
(E) inferior rectal vessels

43. All of the following statements concerning the pectinate line are correct EXCEPT

(A) It is the junction between the superior and inferior parts of the anal canal.
(B) The superior rectal artery supplies the superior part of the anal canal.
(C) Lymphatic vessels from the superior part of the anal canal drain into the internal lymph nodes.
(D) The superior part of the anal canal is drained by the internal rectal venous plexus which drains into tributaries of the caval venous system.
(E) The nerve supply to the anal canal superior to the pectinate line is somatic innervation.

44. The navicular fossa is located in which of the following structures?

(A) glans penis
(B) prostate
(C) bladder
(D) seminal vesicle
(E) expanded in the clitoris

45. Which of the following nerves do NOT innervate the scrotum?

(A) ilioinguinal
(B) genitofemoral
(C) pudendal
(D) posterior femoral cutaneous
(E) obturator

46. The deep arteries of the penis are located in which of the following areas?

(A) superficial to the tunica albuginea
(B) superficial to the deep fascia
(C) within the corpus spongiosum
(D) within the corpora cavernosa
(E) prepuce

47. Which of the following muscles surrounds the crura in the root of the penis?

(A) superficial transverse perineal
(B) deep transverse perineal
(C) bulbospongiosus
(D) ischiocavernosus
(E) cremaster

48. The space between the labia minora, the vestibule, contains all of the following structures EXCEPT

(A) urethral opening
(B) opening of the vagina
(C) ducts of the greater vestibular glands
(D) ducts of the lesser vestibular glands
(E) bulbs of the vestibule

49. Each of the following statements relating to the labia minora is correct EXCEPT

(A) They are folds of fat.
(B) They are hairless.
(C) They are enclosed in the pudendal cleft within the labia majora.
(D) They have a core of spongy connective tissue containing erectile tissue.
(E) They contain many sebaceous glands and sensory nerve endings.

50. Parasympathetic stimulation in the female produces which of the following?

(A) decreased vaginal secretions
(B) engorgement of erectile tissue in the bulbs of the vestibule
(C) engorgement of the greater vestibular gland
(D) decrease in size of the clitoris
(E) the clitoris becomes flaccid

51. The paramesonephric ducts in the male degenerate except for which of the following structures?

(A) efferent ductules of the testis
(B) appendix testis
(C) ductus epididymis
(D) seminal vesicles
(E) ductus deferens

52. The only parts remaining from the mesonephric system in the female include which of the following structures?

(A) fimbriae
(B) uterine tubes
(C) epoöphoron
(D) uterus
(E) cervix

53. The vaginal fornices are derived from which of the following structures?

(A) paramesonephric duct
(B) mesonephric ducts
(C) ectodermal duct
(D) sinovaginal bulbs
(E) urogenital sinus

54. Rapid elongation of the genital tubercle in the male gives rise to which of the following structures?

(A) testes
(B) scrotum
(C) ductus deferens
(D) phallus
(E) epididymis

55. The scrotum arises from which of the following structures?

(A) mesonephric ducts
(B) paramesonephric ducts
(C) urethral folds
(D) cloacal fold
(E) genital swellings

56. The clitoris is derived from which of the following structures?

(A) urethral folds
(B) genital swellings
(C) urogenital groove
(D) mesonephric ducts
(E) genital tubercle

57. Which of the following structures give rise to the labia minora?

(A) mesonephric ducts
(B) urogenital groove
(C) genital swellings
(D) urethral folds
(E) genital tubercle

58. Which of the following structures extends into the labia majora?

(A) suspensory ligament of the ovary
(B) ligament of the ovary proper
(C) processus vaginalis
(D) round ligament of the uterus
(E) uterine tube

59. Which of the following terms correctly applies to the pelvis of the normal female?

(A) spondyloid
(B) android
(C) gynecoid
(D) anthropoid
(E) platypelloid

60. Persons with spondylolysis have a defect in which of the following structures?

(A) vertebral arch
(B) body of L5 vertebra
(C) sacrum
(D) vertebral canal
(E) vertebral body

61. The neck of the bladder in females is held firmly by which of the following structures?

(A) puboprostatic ligaments
(B) puborectalis muscle
(C) levator ani muscle
(D) coccygeus muscle
(E) pubovesical ligaments

62. The median umbilical ligament contains which of the following structures?

(A) inferior epigastric vessels
(B) obturator vessels
(C) remnant of the urachus
(D) uterine tubes
(E) ovarian vessels

63. The inferior ends of the seminal vesicles are closely related to the rectum and are separated from it only by which of the following structures?

(A) pubovesical ligament
(B) puboprostatic ligament
(C) rectovesical septum
(D) puborectalis muscle
(E) coccygeus muscle

64. Which of the following structures is the largest accessory gland of the male reproductive system?

(A) testes
(B) seminal vesicles
(C) prostate
(D) bulbourethral glands
(E) epididymis

65. All of the following muscles compress the vagina and act like sphincters EXCEPT

(A) pubovaginalis
(B) external urethral sphincter
(C) urethrovaginal sphincter
(D) bulbospongiosus
(E) pubococcygeus

66. The mesosalpinx forms the mesentery for which of the following structures?

(A) uterine tube
(B) ovary
(C) small intestine
(D) bladder
(E) uterus

67. The ovarian arteries arise from which of the following arteries?

(A) superior gluteal
(B) inferior gluteal
(C) abdominal aorta
(D) superior rectal
(E) internal pudendal

68. Which of the following structures provides support for the ampulla of the rectum?

(A) urogenital diaphragm
(B) puboprostatic ligament
(C) sacrogenital ligament
(D) tendinous arch of pelvic fascia
(E) levator ani and anococcygeal ligament

69. In males the rectum is related anteriorly to all of the following structures EXCEPT

(A) fundus of the urinary bladder
(B) proximal parts of the ureters
(C) ductus deferens
(D) seminal vesicles
(E) prostate

70. Some obstetricians apply the term *perineum* to a more restricted region that extends between which of the following structures?

(A) perineal body and the mons pubis
(B) vagina and anus
(C) pubic arch and the rectum
(D) ischial spines and pubic tubercles
(E) vestibule and sacrum

71. Distally the corpus spongiosum expands to form which of the following structures?

(A) bulb of the penis
(B) clitoris
(C) vestibule
(D) glans penis
(E) crura of the penis

72. Helicine arteries are located in which of the following structures?

(A) superficial perineal space
(B) deep perineal space
(C) corpora cavernosa
(D) spermatic cord
(E) bulbospongiosum

73. The term *vulva* is synonymous with which of the following structures?

(A) mons pubis
(B) labia majora
(C) vestibule
(D) clitoris
(E) pudendum

74. Which of the following structures may be referred to as the fourchette?

(A) prepuce of the clitoris
(B) frenulum of the labia minora
(C) vestibule of the vagina
(D) glans clitoris
(E) mons pubis

75. The lesser vestibular glands open into which of the following structures?

(A) vestibule between the urethral and vaginal orifices
(B) vestibule on each side of the vaginal orifice
(C) bulbs of the vestibule
(D) glans clitoris
(E) bulbourethral ducts

76. Which of the following structures is incised during median episiotomy for childbirth?

(A) perineal body
(B) greater vestibular glands
(C) lesser vestibular glands
(D) clitoris
(E) urethra

77. Which of the following arteries is a direct continuation of the inferior mesenteric artery?

(A) superior rectal
(B) middle rectal
(C) inferior rectal
(D) iliolumbar
(E) lateral sacral

78. The superior gluteal artery leaves the pelvis through which of the following openings?

(A) greater sciatic foramen
(B) lesser sciatic foramen
(C) obturator canal
(D) pudendal canal
(E) sacral foramina

79. Which of the following structures separates the bladder from the pubic bones in females?

(A) rectouterine fold
(B) vesicouterine pouch
(C) trigone of the bladder
(D) median umbilical fold
(E) retropubic space

80. The membranous part of the male urethra is located in which of the following structures?

(A) bladder
(B) prostate
(C) external urethral sphincter
(D) bulb of penis
(E) glans penis

81. The paraurethral glands are homologues to which of the following structures?

(A) testes
(B) prostate
(C) seminal vesicles
(D) bulbourethral glands
(E) epididymis

82. Lithotripsy uses shock waves to break up which of the following structures?

(A) blood clots
(B) foreign bodies
(C) small tumors
(D) urinary calculi
(E) abscesses

83. Which of the following areas contributes to the major part of the prostate?

(A) anterior lobe
(B) isthmus
(C) posterior lobe
(D) lateral lobes
(E) middle lobe

84. The relationship ("water passing under the bridge") is an especially important one for surgeons ligating which of the following arteries?

(A) ovarian
(B) testicular
(C) uterine
(D) vaginal
(E) obturator

85. Immediately superior to the perineal membrane is located which of the following muscles?

(A) obturator internus
(B) levator ani
(C) bulbocavernosus
(D) ischiocavernous
(E) deep transverse perineal

86. The navicular fossa is located in which of the following structures?

(A) bulb of the penis
(B) prostate
(C) membranous urethra
(D) glans penis
(E) neck of bladder

87. All of the following nerves provide innervation to the scrotum EXCEPT

(A) obturator
(B) ilioinguinal
(C) genitofemoral
(D) pudendal
(E) posterior femoral cutaneous

88. All of the following structures surround the corpora cavernosa and corpus spongiosum EXCEPT

(A) loose areolar tissue
(B) deep fascia
(C) tunica albuginea
(D) skin
(E) tunica dartos

89. Which of the following arteries gives rise to the deferential artery?

(A) internal pudendal
(B) testicular
(C) inferior vesicle
(D) obturator
(E) umbilical

90. Which of the following structures is located at the free anterior borders of the levator ani?

(A) coccygeus muscle
(B) piriformis muscle
(C) urogenital hiatus
(D) obturator internus muscle
(E) rectum

91. Which of the following structures form a U-shaped sling around the anorectal junction?

(A) sacrospinous ligament
(B) anococcygeal ligament
(C) superficial transverse perineal muscle
(D) puborectalis muscle
(E) sacrotuberous ligament

92. Which of the following muscles is the larger part and most important muscle in the pelvic floor?

(A) coccygeus
(B) obturator internus
(C) piriformis
(D) deep transverse perineal muscle
(E) levator ani

93. Which of the following muscles passes through the lesser sciatic foramen?

(A) obturator internus
(B) piriformis
(C) puborectalis
(D) pubococcygeus
(E) iliococcygeus

94. Which of the following muscles leaves the lesser pelvis through the greater sciatic foramen?

(A) pubococcygeus
(B) iliococcygeus
(C) puborectalis
(D) piriformis
(E) coccygeus

95. The pelvic splanchnic nerves contain which of the following?

(A) somatic afferents
(B) sympathetic fibers
(C) parasympathetic fibers
(D) special visceral efferents
(E) special somatic afferents

96. The internal iliac artery is crossed by which of the following structures at the level of the 4th intervertebral disc between L5 and S1?

(A) puborectalis muscle
(B) ureter
(C) obturator nerve
(D) umbilical artery
(E) tendinous arch of the levator ani

97. The superior end of the vagina surrounds which of the following structures?

(A) round ligament
(B) urogenital hiatus
(C) urethra
(D) cervix
(E) neck of bladder

98. Which of the following nerves innervates the lower one fourth of the vagina?

(A) pelvic splanchnics
(B) lumbar splanchnics
(C) pudendal
(D) superior hypogastric plexus
(E) inferior hypogastric plexus

99. Which of the following structures forms the mesentery of the uterus?

(A) round ligament of the uterus
(B) mesosalpinx
(C) pelvic diaphragm
(D) endopelvic fascia
(E) mesometrium

100. The posterior part of the tendinous arch of pelvic fascia forms which of the following ligaments?

(A) puboprostatic
(B) pubovesicle
(C) transverse cervical
(D) sacrogenital
(E) sacrospinous

DIRECTIONS (Questions 101 through 105): Identify the anatomical features indicated on the art below.

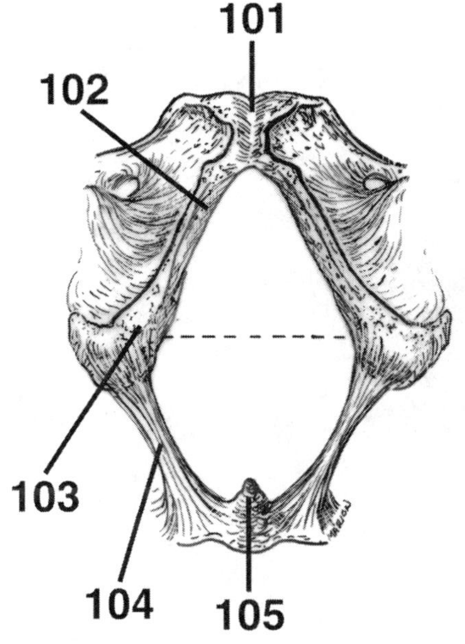

DIRECTIONS (Questions 106 through 110): Identify the anatomical features indicated on the art below.

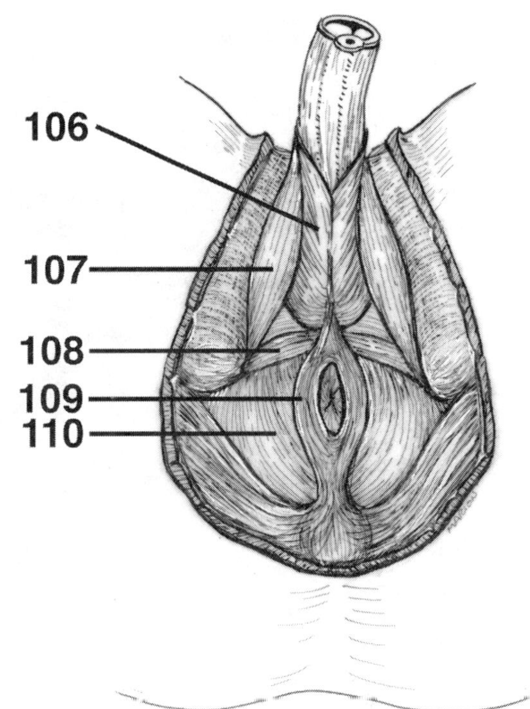

DIRECTIONS (Questions 111 through 115): Identify the anatomical features indicated on the art below.

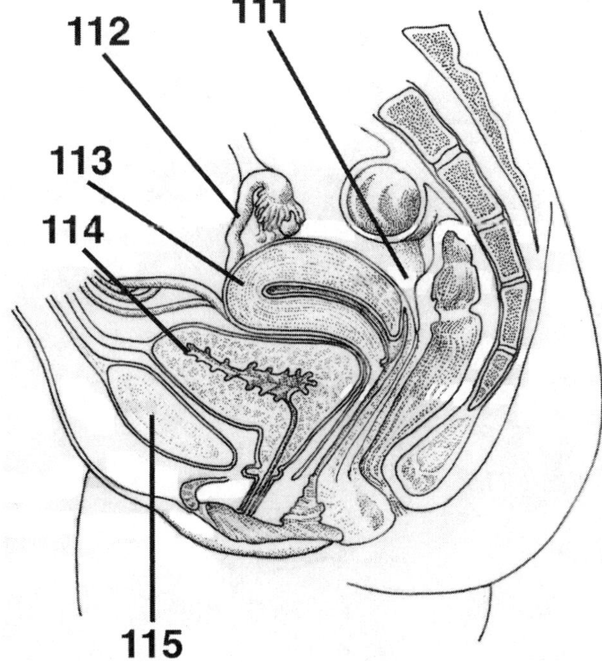

DIRECTIONS (Questions 116 through 120): Identify the anatomical features indicated on the art below.

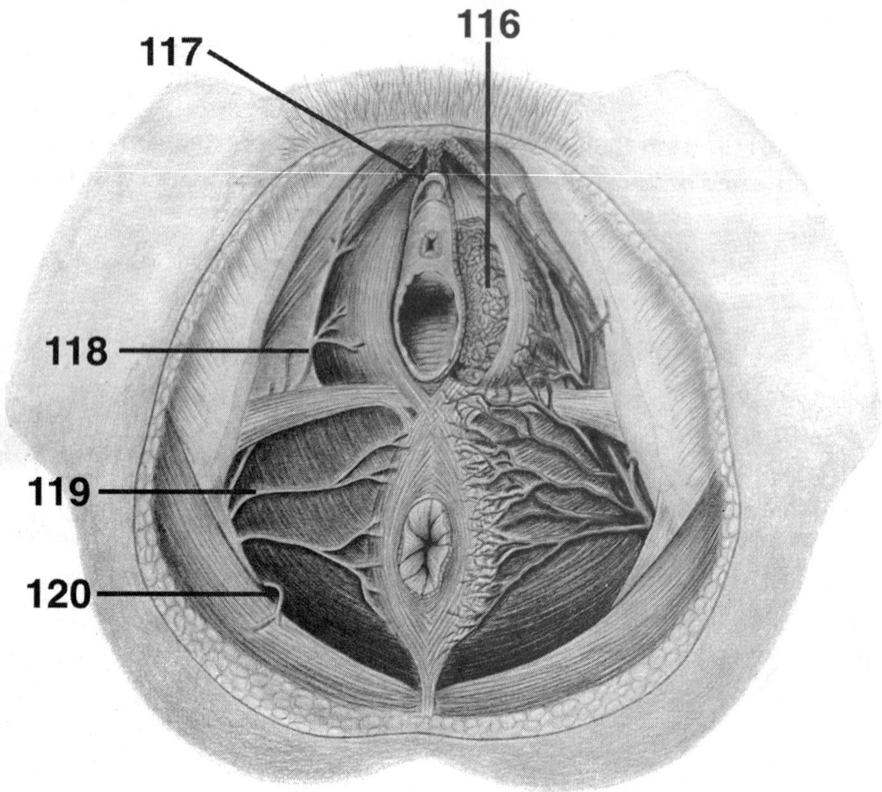

DIRECTIONS (Questions 121 through 125): Identify the anatomical features indicated on the art below.

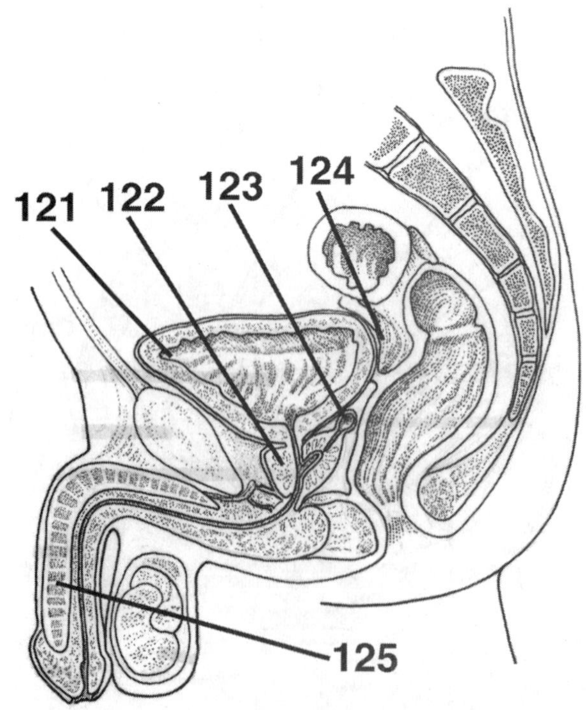

Answers and Explanations

1. **(E)** The superior boundary of the pelvic cavity is the pelvic inlet. The pelvis is limited inferiorly by the pelvic outlet, which is closed by the musculofascial pelvic diaphragm and bounded posteriorly by the coccyx and anteriorly by the pubic symphysis *(Moore, p 333).*

2. **(D)** The pelvic girdle is formed by the hip (ischium, pubis, and ilium) bones and the sacrum *(Moore, p 333).*

3. **(B)** The concavity between the ischial spine and the ischial tuberosity is the lesser sciatic notch. The larger concavity, the greater sciatic notch, is superior to the ischial spine and is formed in part by the ilium *(Moore, p 334).*

4. **(C)** The pubis is an angulated bone with a superior ramus that helps to form the acetabulum and an inferior ramus that helps to form the obturator foramen. A thickening on the anterior part of the body of the pubis is the pubic crest, which ends laterally as a prominent bump, the pubic tubercle. The lateral part of the superior ramus has an oblique ridge known as the pecten pubis or pectineal line of the pubis *(Moore, p 335).*

5. **(A)** The pelvic inlet is bounded by the superior margin of the pubic symphysis anteriorly, posterior border of the pubic crest, pecten pubis, arcuate line of the ilium, anterior border of the ala of the sacrum, and the sacral promontory *(Moore, p 336).*

6. **(B)** When a person is in the anatomical position, the anterior superior iliac spines and the anterior aspect of the pubic symphysis lie in the same vertical plane *(Moore, p 337).*

7. **(A)** Weak areas of the pelvis include the pubic rami, acetabula, sacroiliac joint, and alae of the ilium *(Moore, p 338).*

8. **(C)** The piriformis muscles cover the posterolateral wall of the pelvis. The pelvic floor is formed by the funnel-shaped pelvic diaphragm, which consists of the levator ani and coccygeus. The levator ani includes the pubococcygeus, puborectalis, and iliococcygeus *(Moore, p 342).*

9. **(A)** The bony framework of the lateral pelvic walls is formed by the hip bones and the obturator foramen, which is closed by the obturator membrane. The obturator internus muscles cover and thus pad most of the lateral pelvic walls *(Moore, p 341).*

10. **(B)** The two main nerves of the sacral plexus include the sciatic and pudendal. The sciatic nerve is the largest and broadest nerve in the body. It is formed by the ventral rami of L4-S3, which converge on the anterior surface of the piriformis. The sciatic nerve usually passes through the greater sciatic foramen, inferior to the piriformis, to enter the gluteal region *(Moore, p 347).*

11. **(D)** The pudendal nerve is derived from the anterior divisions of the ventral rami of S2 through S4. It accompanies the internal pudendal artery and leaves the pelvis through the greater sciatic foramen between the piriformis and coccygeus muscles. It is the main nerve of the perineum and the chief sensory nerve of the external genitalia *(Moore, p 347).*

12. **(C)** The superior gluteal nerve arises from the posterior divisions of the ventral rami of L4

through S1 and leaves the pelvis through the greater sciatic foramen, superior to the piriformis *(Moore, p 347).*

13. **(A)** The sacral sympathetic trunks descend posterior to the rectum in the extraperitoneal connective tissue and send gray rami communicantes to each of the ventral rami of the sacral and coccygeal nerves. Each of the sacral trunks is smaller than the lumbar trunks and usually has four sympathetic ganglia *(Moore, p 350).*

14. **(D)** The pelvic splanchnic nerves merge with the hypogastric nerves to form the inferior hypogastric (and pelvic) plexuses. *(Moore, p 350).*

15. **(E)** Four main arteries enter the lesser or true pelvis. The internal iliac and ovarian arteries are paired, and the median sacral and superior rectal arteries are unpaired. The testicular artery does not enter the true pelvis, as it follows the testes through the inguinal canal into the scrotum *(Moore, p 350).*

16. **(D)** The internal iliac artery is the artery of the pelvis; however, it does supply branches to the buttocks, thighs, and the perineum *(Moore, p 350).*

17. **(E)** The superior gluteal and iliolumbar arteries arise from the posterior division of the internal iliac arteries. The gonadal arteries arise from the abdominal aorta and the superior rectal arteries from the inferior mesenteric artery. The umbilical, obturator, uterine, vaginal, internal pudendal, and inferior gluteal arteries arise from the anterior division of the internal iliac arteries *(Moore, pp 354–355).*

18. **(C)** The ureters are muscular tubes, 25 to 30 cm long, that connect the kidneys to the urinary bladder. They are retroperitoneal, with their superior halves in the abdomen and their inferior halves in the pelvis. The pelvic part of the ureters begins where it crosses the bifurcation of the common iliac artery *(Moore, p 357).*

19. **(B)** The uvula of the bladder is a slight projection of the trigone of the bladder. It is usually more prominent in older men *(Moore, p 362).*

20. **(A)** Parasympathetic fibers to the bladder are derived from the pelvic splanchnic nerves. They are motor to the detrusor muscle and inhibitory to the internal sphincter. Sympathetic fibers to the bladder are derived from T11 through L2 *(Moore, p 362).*

21. **(B)** The prostatic urethra is the widest and most dilatable part of the male urethra *(Moore, p 363).*

22. **(C)** The internal surface of the posterior wall of the prostatic urethra has a median ridge known as the urethral crest. A groove on each side, known as the prostatic sinus, receives the prostatic ductules. A rounded eminence located in the middle of the median ridge is known as the seminal colliculus. The prostatic utricle is an embryonic remnant of the uterus and part of the vagina. The ejaculatory ducts open beside the prostatic utricle *(Moore, p 364).*

23. **(A)** The paraurethral glands are homologues to the prostate. They have common paraurethral ducts, which open, one on each side, near the external urethral orifice *(Moore, p 364).*

24. **(A)** The ductus deferens begins in the tail of the epididymis and ascends in the spermatic cord. It passes through the inguinal canal and enters the pelvis. It ends by joining the duct of the seminal vesicle to form the ejaculatory duct. It descends medial to the ureter and seminal vesicle *(Moore, p 367).*

25. **(E)** Ducts of the bulbourethral glands empty into the urethra. The seminal vesicles are elongated structures that lie between the fundus of the bladder and the rectum, and they do not store sperm. They secrete a thick, alkaline fluid that mixes with the sperm as they pass into the ejaculatory ducts to the urethra. The superior ends of the seminal vesicles are covered with peritoneum and lie posterior to the ureters, where the rectovesical pouch separates them from the rectum. The inferior ends of the seminal vesicle are separated from the rectum by the rectovesical septum *(Moore, p 368).*

26. **(B)** The posterior fornix is the deepest part of the vagina and is closely related to the rectouterine pouch. The vaginal fornix is the recess around the cervix and is described as having anterior, posterior, and lateral parts *(Moore, pp 372–373).*

27. **(C)** The uterus is a thick-walled, pear-shaped muscular organ whose body lies between the layers of the broad ligament. In the adult, the uterus is usually anteverted and anteflexed; its position changes with the degree of fullness of the bladder and rectum *(Moore, pp 373–374).*

28. **(E)** Only the cylindrical, narrow inferior part of the uterus known as the cervix protrudes into the uppermost vagina *(Moore, p 376).*

29. **(B)** The rounded vaginal part of the cervix extends into the vagina and communicates with it through the external os. The cervical canal is broadest at its middle part and communicates with the uterine cavity through the internal os and with the vagina through the external os *(Moore, pp 376–377).*

30. **(D)** Laterally, the peritoneum of the broad ligament is prolonged superiorly over the vessels as the suspensory ligament of the ovary. The part of the broad ligament by which the ovary is suspended is the mesovarium. The part of the broad ligament forming the mesentery of the uterine tube is the mesosalpinx. The major part of the broad ligament, the mesentery of the uterus, or mesometrium, is below the mesosalpinx and mesovarium *(Moore, p 381).*

31. **(B)** The ampulla is the widest and longest part of the uterine tubes. It begins at the medial end of the infundibulum. Oocytes expelled from the ovaries are usually fertilized in the ampulla *(Moore, p 383).*

32. **(C)** The distal end of the ovary connects to the lateral wall of the pelvis by the suspensory ligament of the ovary. The ligament conveys the ovarian vessels, lymphatics, and nerves to and from the ovary and constitutes the lateral part of the mesovarium of the broad ligament. The ovary also attaches to the uterus by the ligament of the ovary *(Moore, p 384).*

33. **(C)** The parasympathetic fibers in the ovarian plexus are derived from the vagus nerve *(Moore, p 384).*

34. **(A)** The rectosigmoid junction lies anterior to the S3 vertebra *(Moore, p 384).*

35. **(C)** The osseofibrous structures marking the boundaries of the perineum include the pubic symphysis, inferior pubic rami, ischial rami, ischial tuberosities, sacrotuberous ligaments, inferiormost sacrum, and coccyx *(Moore, p 389).*

36. **(A)** A transverse line joining the anterior ends of the ischial tuberosities divides the perineum into two triangles. The anal triangle, containing the anus, is posterior to this line. The urogenital triangle, containing the root of the scrotum and penis in males and the external genitalia in females is anterior to this line *(Moore, pp 389–390).*

37. **(A)** The bulbospongiosus, external anal sphincter, and superficial and deep transverse perineal muscles converge at the site of the perineal body *(Moore, p 390).*

38. **(E)** The following structures are found in the superficial perineal space: the root of the penis and its associated muscles, the ischiocavernous and bulbospongiosus. The proximal part of the spongy urethra, superficial transverse perineal muscles, internal pudendal vessels, and branches of the pudendal nerve are also located in the superficial perineal space *(Moore, p 394).*

39. **(E)** In females, the deep perineal pouch contains the proximal part of the urethra, the external urethral sphincter muscle, the deep transverse perineal muscles, and related vessels and nerves *(Moore, p 394).*

40. **(C)** The pudendal canal is a space within the obturator fascia, which covers the medial aspect of the obturator internus and lines the lateral wall of the ischioanal fossa *(Moore, p 395).*

41. **(E)** The pudendal nerve supplies most of the innervation to the perineum. Toward the distal end of the pudendal canal, the pudendal nerve

splits, giving rise to the perineal nerves and continuing as the dorsal nerve of the penis or clitoris *(Moore, p 395).*

42. **(C)** The superior half of the anal canal contains mucous membrane that is characterized by a series of longitudinal ridges called anal columns. These columns contain the terminal branches of the superior rectal artery and vein *(Moore, p 395).*

43. **(D)** The pectinate line indicates the junction of the superior part of the anal canal and the inferior part. The anal canal superior to the pectinate line differs from the part inferior to the pectinate line in its arterial supply, innervation, and venous and lymphatic drainage. The superior rectal artery supplies the superior part of the anal canal and the inferior rectal arteries supply the inferior part. Superior to the pectinate line, the internal rectal venous plexus drains chiefly into the superior rectal vein—a tributary of the inferior mesenteric vein and the portal system. Inferior to the pectinate line the internal rectal plexus drains into the inferior rectal veins—tributaries to the caval venous system. Superior to the pectinate line, the lymphatics drain into the internal iliac lymph nodes into the common iliac and lumbar nodes. Inferior to the pectinate line, the lymphatics drain into the superficial inguinal lymph nodes. Superior to the pectinate line, the innervation is visceral; inferior to the pectinate line, the innervation is somatic *(Moore, pp 398–400).*

44. **(A)** The lumen of the spongy urethra is approximately 5 mm in diameter and expanded in the bulb of the penis to form the intrabulbar fossa and in the glans penis to form the fossa navicularis *(Moore, p 403).*

45. **(E)** The anterior aspect of the scrotum is supplied by anterior scrotal nerves derived from the ilioinguinal nerve and by the genital branch of the genitofemoral nerve. The posterior aspect of the scrotum is supplied by posterior scrotal nerves, branches of the superficial perineal nerves, and the perineal branch of the posterior femoral cutaneous nerve *(Moore, p 405).*

46. **(D)** The arterial supply of the penis is mainly from the branches of the internal pudendal arteries. The deep arteries of the penis are the main vessels supplying the cavernous spaces in the erectile tissue of the corpora cavernosa and are therefore involved in the erection of the penis. The deep arteries pierce the crura and run within the corpora cavernosa *(Moore, p 410).*

47. **(D)** The ischiocavernous muscles surround the crura in the root of the penis. Each muscle arises from the internal surface of the ischial tuberosity and ischial ramus and passes anteriorly on the crus of the penis, where it is inserted into the sides and ventral surface of the crus and the perineal membrane. The ischiocavernous muscles force blood from the cavernous spaces in the crura into the distal parts of the corpora cavernosa, thus increasing the turgidity of the penis *(Moore, p 409).*

48. **(E)** The vestibule is the space between the labia minora containing the openings of the urethra, vagina, and ducts of the greater and lesser vestibular glands. The greater vestibular glands are round or oval and are partly overlapped posteriorly by bulbs of the vestibule and, like the bulbs, are partially surrounded by the bulbospongiosus muscles *(Moore, p 413).*

49. **(A)** The labia minora are folds of fat-free, hairless skin. They are enclosed in the pudendal cleft within the labia majora, immediately surrounding the vestibule of the vagina. They have a core of spongy connective tissue containing erectile tissue and many small blood vessels *(Moore, p 413).*

50. **(B)** Parasympathetic stimulation in the female produces an increase in vaginal secretions, erection of the clitoris, and engorgement of erectile tissue in the bulbs of the vestibule *(Moore, p 415).*

51. **(B)** Except for the most cranial portion, the appendix epididymis, the mesonephric ducts persist and form the main genital ducts. Immediately below the entrance of the efferent ductules, the mesonephric ducts elongate and become highly convoluted, forming the ductus

epididymis. From the tail of the epididymis to the outbudding of the seminal vesicle, the mesonephric ducts obtain a thick muscular coat and form the ductus deferens. The region of the ducts beyond the seminal vesicles is the ejaculatory duct. Except for a small portion at their cranial ends, the appendix testis, the paramesonephric ducts in the male degenerate *(Sadler, p 326).*

52. **(C)** The only parts remaining from the mesonephric system are the epoöphoron, paroöphoron, and Gartner's cyst *(Sadler, p 325).*

53. **(A)** The winglike expansions of the vagina around the end of the uterus, the vaginal fornices, are of paramesonephric origin *(Sadler, p 329).*

54. **(D)** The rapid elongation of the genital tubercle in the male gives rise to the phallus *(Sadler, p 331).*

55. **(E)** The genital swellings, known in the male as the scrotal swellings, arise in the inguinal region. With further development, they move caudally, and each swelling then makes up half of the scrotum *(Sadler, p 332).*

56. **(E)** In females, the genital tubercle elongates only slightly and forms the clitoris *(Sadler, p 335).*

57. **(D)** The urethral folds do not fuse, as in the male, but develop into the labia minora *(Sadler, p 336).*

58. **(D)** Descent of the gonads is considerably less in the female than in the male, and the ovaries finally settle just below the rim of the true pelvis. The cranial genital ligament forms the suspensory ligament of the ovary, whereas the caudal genital ligament forms the ligament of the ovary proper and the round ligament of the uterus. The latter extends into the labia majora *(Sadler, p 342).*

59. **(C)** Android and anthropoid pelves are common in males. The platypelloid pelvis is uncommon in both males and females. The gynecoid pelvis is the normal female type of pelvis *(Moore, p 337).*

60. **(A)** Persons with spondylolysis have a defect in the vertebral arch. When this is bilateral, it results in the L5 vertebra being divided into two pieces. If the parts separate, the abnormality is spondylolisthesis, which is anterior displacement of the body of the L5 vertebra on the sacrum *(Moore, p 339).*

61. **(E)** The bladder is relatively free within the extraperitoneal subcutaneous fatty tissue except for its neck, which is held firmly by the puboprostatic ligaments in males and the pubovesical ligaments in females *(Moore, p 359).*

62. **(C)** The median umbilical ligament is formed by the remnant of the urachus *(Moore, p 361).*

63. **(C)** The inferior ends of the seminal vesicles are closely related to the rectum and are separated from it only by the rectovesical septum, a membranous partition *(Moore, p 368).*

64. **(C)** The prostate is the largest accessory gland of the male reproductive system *(Moore, p 369).*

65. **(E)** The pubovaginalis, external urethral sphincter, urethrovaginal sphincter, and bulbospongiosus compress the vagina and act like sphincters *(Moore, p 372).*

66. **(A)** The part of the broad ligament forming the mesentery of the uterine tube is the mesosalpinx *(Moore, p 377).*

67. **(C)** The ovarian arteries arise from the abdominal aorta *(Moore, p 384).*

68. **(E)** The dilated terminal part of the rectum, lying directly above and supported by the pelvic diaphragm (levator ani) and anococcygeal ligament is the ampulla of the rectum *(Moore, p 385).*

69. **(B)** In males the rectum is related anteriorly to the fundus of the urinary bladder, terminal parts of the ureters, ductus deferens, seminal vesicles, and prostate *(Moore, p 385).*

70. **(B)** Some obstetricians apply the term *perineum* to a more restricted region that includes the area between the vagina and anus *(Moore, p 389).*

71. **(D)** Distally the corpus spongiosum expands to form the conical glans penis *(Moore, p 407).*

72. **(C)** The deep arteries of the penis are the main vessels supplying the cavernous spaces in the erectile tissue of the corpora cavernosa and are therefore involved in the erection of the penis. They give off numerous branches that open directly into the cavernous spaces. When the penis is flaccid, these arteries are coiled and therefore are called helicine arteries *(Moore, p 410).*

73. **(E)** The synonymous terms *vulva* and *pudendum* include the clitoris, vestibule of the vagina, bulbs of vestibule, and greater vestibular gland *(Moore, p 413).*

74. **(B)** In young women, especially virgins, the labia minora are connected by a small fold known as the frenulum of the labia minora or the fourchette *(Moore, p 413).*

75. **(A)** The slender ducts of the greater vestibular glands pass deep to the bulbs of the vestibule and open into the vestibule on each side of the vaginal orifice. The lesser vestibular glands are small glands on each side of the vestibule that open into it between the urethral and vaginal orifices *(Moore, p 414).*

76. **(A)** The perineal body is the major structure incised during median episiotomy for childbirth *(Moore, p 391).*

77. **(A)** The superior rectal artery is the direct continuation of the inferior mesenteric artery. The superior rectal artery anastomoses with branches of the middle rectal artery (a branch of the internal iliac artery) and with the inferior rectal artery (a branch of the internal pudendal artery) *(Moore, p 355).*

78. **(B)** The superior gluteal artery leaves the pelvis through the superior part of the greater sciatic foramen, superior to the piriformis muscle, to supply the gluteal muscles in the buttocks *(Moore, p 355).*

79. **(E)** When empty, the adult male or female urinary bladder is in the lesser pelvis, lying posterior and slightly superior to the pubic bones. It is separated from these bones by the potential retropubic space and lies inferior to the peritoneum, where it rests on the pelvic floor *(Moore, p 359).*

80. **(C)** The intermediate part of the urethra (membranous part) is the section passing through the external urethral sphincter and the perineal membrane. The short intermediate part, extending from the prostatic urethra to the spongy urethra, is the narrowest and least distensible part of the urethra *(Moore, p 364).*

81. **(B)** Urethral glands are present particularly in the superior part of the female urethra. The paraurethral glands are homologues to the prostate *(Moore, p 364).*

82. **(D)** Lithotripsy uses shock waves to break up a stone into small fragments that can be passed in the urine *(Moore, p 358).*

83. **(D)** The lateral lobes on either side of the urethra form the major part of the prostate *(Moore, p 369).*

84. **(C)** In its uppermost portion, at the base of the peritoneal broad ligament, the uterine artery runs transversely toward the cervix while the ureters pass immediately beneath them as they pass on each side of the cervix toward the bladder. This relationship ("water passing under the bridge") is an especially important one for surgeons ligating the uterine artery, as in a hysterectomy *(Moore, p 380).*

85. **(E)** Immediately superior to the perineal membrane is the deep transverse perineal muscle *(Moore, p 390).*

86. **(D)** The lumen of the spongy urethra is approximately 5 mm in diameter: however, it is expanded in the bulb of the penis to form the

intrabulbar fossa and in the glans penis to form the navicular fossa *(Moore, p 403).*

87. **(A)** The anterior aspect of the scrotum is supplied by anterior scrotal nerves derived from the ilioinguinal and the genital branch of the genitofemoral nerve. The posterior aspect of the scrotum is supplied by posterior scrotal nerves, provided by perineal branches of the pudendal and perineal branches of the posterior femoral cutaneous nerve *(Moore, p 405).*

88. **(E)** The penis is composed of three cylindrical bodies of erectile cavernous tissue enclosed by a fibrous capsule, the tunica albuginea. Superficial to the capsule is the deep fascia of the penis (Buck's fascia). Superficial to the deep fascia is the loose areolar tissue, which lies just beneath the skin of the penis. The tunica dartos is located in the scrotum *(Moore, p 406).*

89. **(C)** The tiny deferential artery usually arises from the inferior vesical artery and terminates by anastomosing with the testicular artery, posterior to the testis *(Moore, p 367).*

90. **(C)** The free anterior borders of the levator ani are separated by a gap, the urogenital hiatus, through which the urethra (and, in the female, the vagina) passes *(Moore, p 357).*

91. **(D)** The puborectalis muscle unites with its partner to from a U-shaped sling around the anorectal junction. The puborectalis is responsible for the anorectal angle (perineal flexure), which is important in maintaining fecal continence *(Moore, p 345).*

92. **(E)** The levator ani, a broad muscular sheet, is the largest and most important muscle in the pelvic floor *(Moore, p 341).*

93. **(A)** Each obturator internus passes posteriorly from the lesser pelvis through the lesser sciatic foramen and turns sharply laterally to attach to the greater trochanter of the femur *(Moore, p 341).*

94. **(D)** The piriformis muscle leaves the lesser pelvis through the greater sciatic foramen to attach to the upper border of the greater trochanter of the femur *(Moore, p 341).*

95. **(C)** The pelvic splanchnic nerves contain parasympathetic fibers derived from S2, S3, and S4 spinal cord segments and visceral afferent fibers from cell bodies in the spinal ganglia of the corresponding spinal nerves *(Moore, p 350).*

96. **(B)** The internal iliac artery begins at the level of the fourth disc between L5 and S1 vertebrae, where it is crossed by the ureter. It is separated from the sacroiliac joint by the internal iliac vein and the lumbosacral trunk *(Moore, p 350).*

97. **(D)** The superior end of the vagina surrounds the cervix; the lower end passes anteroinferiorly through the pelvic floor to open in the vestibule *(Moore, p 371).*

98. **(C)** Only the lower one-fifth to one-fourth of the vagina is somatic in terms of innervation. The innervation of this part of the vagina is from the deep perineal branch of the pudendal nerve *(Moore, p 373).*

99. **(E)** The major part of the broad ligament, the mesentery of the uterus or mesometrium, is below the mesosalpinx and mesovarium *(Moore, p 377).*

100. **(D)** The anterior part of the tendinous arch of pelvic fascia forms the puboprostatic ligament in males or the pubovesical ligament in females. The posterior part of the tendinous arch of pelvic fascia forms the sacrogenital ligaments from the sacrum around the side of the rectum to attach to the prostate in the male or the vagina in the female *(Moore, p 380).*

101. pubic symphysis

102. inferior pubic ramus

103. ischial tuberosity

104. sacrotuberous ligament

105. coccyx

106. bulbocavernosus

107. ischiocavernous

108. superficial transverse perineal muscle

109. anal sphincter

110. levator ani

111. rectouterine pouch

112. uterine tube

113. fundus of uterus

114. apex of bladder

115. pubic symphysis

116. bulb of vestibule

117. prepuce of clitoris

118. perineal branches of pudendal nerve

119. inferior rectal nerve

120. inferior cluneal nerve

121. fundus of bladder

122. prostate gland

123. seminal vesicle

124. rectovesical pouch

125. corpora cavernosa

CHAPTER 6

The Lower Limb

Questions

DIRECTIONS (Questions 1 through 100): Each of the numbered items or incomplete statements in this section is followed by answers or by completions of the statement. Select the ONE lettered answer or completion that is BEST in each case.

1. All of the following statements concerning the femur are correct EXCEPT

(A) It is the longest bone in the body.
(B) It is the heaviest bone in the body.
(C) Its length is approximately a quarter of the person's height.
(D) The distal end of the femur undergoes ossification just before birth.
(E) Its average adult length is 36 inches.

2. The adductor tubercle is located on which of the following bones?

(A) femur
(B) tibia
(C) fibula
(D) ischium
(E) ilium

3. The margin of the acetabulum is deficient inferiorly at which of the following structures?

(A) ischial tuberosity
(B) obturator foramen
(C) linea aspera
(D) acetabular notch
(E) ischial spine

4. A deficiency in the smooth articular surface of the acetabulum is known as which of the following?

(A) pecten pubis
(B) pubic crest
(C) lunate surface
(D) ala
(E) ramus

5. A fracture of which of the following bones is among the most troublesome and problematic of all fractures?

(A) tibia
(B) fibula
(C) patella
(D) femoral neck
(E) medial malleolus

6. When it is said that an elderly person has a "broken hip," the usual injury is a fracture of which of the following structures?

(A) acetabulum
(B) neck of the femur
(C) pelvic rami
(D) ischial tuberosity
(E) ilium

7. The medial and lateral malleoli articulate with which of the following bones?

(A) femur
(B) calcaneus
(C) talus
(D) cuboid
(E) fibula

8. Which of the following bones is the most common site for a compound fracture?

(A) femur
(B) tibia
(C) fibula
(D) ilium
(E) ischium

9. Which of the following bones is the largest and strongest bone of the foot?

(A) talus
(B) calcaneus
(C) cuboid
(D) navicular
(E) lateral cuneiform

10. The sustentaculum tali projects from the superior surface of which of the following bones?

(A) intermediate cuneiform
(B) talus
(C) cuboid
(D) navicular
(E) calcaneus

11. Which of the following bones articulates with the fibula, tibia, calcaneus, and navicular?

(A) talus
(B) lateral cuneiform
(C) intermediate cuneiform
(D) medial cuneiform
(E) cuboid

12. Which of the following statements applies to the 2nd metatarsal bone?

(A) It is the shortest.
(B) It is the strongest.
(C) It is the longest.
(D) Its base has a large tuberosity.
(E) Its base articulates with the navicular bone.

13. The deep fascia of the thigh is known as which of the following?

(A) Scarpa's fascia
(B) Colles' fascia
(C) crural fascia
(D) fascia lata
(E) plantar fascia

14. Which of the following structures covers the saphenous opening in the fascia lata?

(A) lateral intermuscular septum
(B) medial intermuscular septum
(C) crural fascia
(D) cribriform fascia
(E) iliotibial tract

15. Which of the following structures passes through the saphenous opening?

(A) femoral artery
(B) femoral vein
(C) femoral nerve
(D) obturator nerve
(E) great saphenous vein

16. The small saphenous vein empties into which of the following veins?

(A) femoral
(B) tibial
(C) peroneal
(D) popliteal
(E) great saphenous

17. Which of the following statements correctly applies to the iliopsoas muscle?

(A) It is a flat quadrangular muscle.
(B) It is the chief flexor of the thigh.
(C) It is enclosed between two layers of fascia lata.
(D) It inserts into the iliotibial tract.
(E) It is located in the posterior compartment of the thigh.

18. All of the following statements concerning the sartorius muscle are correct EXCEPT

(A) It is known as the "tailor's muscle."
(B) It is the longest muscle in the body.
(C) It acts across two joints.
(D) It extends the hip.

(E) It is located in the anterior compartment of the thigh.

19. All of the following statements concerning the quadriceps femoris are correct EXCEPT

(A) Collectively constitutes the largest and most powerful muscle group in the body.
(B) It is the great extensor of the thigh.
(C) It inserts onto the tibia.
(D) It forms the main bulk of the anterior thigh muscles.
(E) The patella provides additional leverage for the quadriceps.

20. Which of the following statements concerning the components of the quadriceps femoris is correct?

(A) The rectus femoris is considered the "kicking muscle."
(B) The vastus lateralis is the smallest component of the quadriceps.
(C) The articularis genus is a derivative of the vastus lateralis.
(D) The rectus femoris lies deep to the vastus intermedius.
(E) The vastus intermedius is the chief flexor of the thigh.

21. Which of the following statements concerning the gracilis muscle is correct?

(A) It passes through the lesser sciatic foramen.
(B) It crosses the knee joint.
(C) It lies deep to the pectineus and adductor longus muscles.
(D) It is located in the anterior compartment of the thigh.
(E) It is a short, fan-shaped muscle.

22. All of the following statements concerning the adductor magnus are correct EXCEPT

(A) It is the largest muscle in the adductor group.
(B) It is located in the posterior compartment of the thigh.
(C) It has adductor and hamstring parts.
(D) It is a composite, triangular muscle with two parts that differ in nerve supply.
(E) Its main action is to adduct the thigh.

23. Which of the following statements correctly applies to the adductor hiatus?

(A) It is an opening in the aponeurotic distal attachment of the adductor longus.
(B) It transmits the femoral nerve, artery and vein.
(C) The opening is located just inferior to the adductor tubercle of the femur.
(D) It extends from the adductor canal in the thigh to the popliteal fossa.
(E) The great saphenous vein passes through the adductor hiatus.

24. All of the following statements concerning the femoral triangle are correct EXCEPT

(A) Its superior border is the inguinal ligament.
(B) Its lateral border is the sartorius.
(C) It is bisected by the femoral artery and vein.
(D) The saphenous nerve passes through the femoral triangle.
(E) Its medial border is the adductor magnus.

25. All of the following statements concerning the femoral sheath are correct EXCEPT

(A) It extends 3 to 4 cm inferior the inguinal ligament.
(B) It is formed by an inferior prolongation of transversalis and iliopsoas fascia.
(C) It encloses the femoral nerve.
(D) Its medial wall is pierced by the great saphenous vein and lymphatic vessels.
(E) It ends by becoming continuous with the adventitia of the femoral vessels.

26. All of the following statements concerning the femoral canal are correct EXCEPT

(A) It contains the femoral nerve.
(B) It is the medial compartment of the femoral sheath.
(C) It allows the femoral vein to expand when venous return from the lower limb is increased.
(D) It extends distally to the level of the proximal edge of the saphenous opening.
(E) It contains loose connective tissue, fat, a few lymphatic vessels, and sometimes a deep inguinal lymph node (Cloquet's node).

27. All of the following statements concerning the femoral ring are correct EXCEPT

(A) The lacunar ligament forms its medial boundary.
(B) The anterior boundary is formed by the inguinal ligament.
(C) The femoral artery forms its lateral boundary.
(D) Its proximal end is closed by extraperitoneal fatty tissue that forms the femoral septum.
(E) It lies anterior to the pectineus muscle.

28. Which of the following statements concerning the femoral artery is correct?

(A) It enters the femoral canal.
(B) It enters the adductor canal.
(C) It passes dorsal to the inguinal ligament.
(D) It gives rise to the inferior epigastric artery.
(E) It passes through the obturator canal.

29. Which of the following statements concerning the deep artery of the thigh is correct?

(A) It is the largest branch of the femoral artery.
(B) It passes through the adductor canal.
(C) It gives rise to the deep circumflex iliac branch.
(D) It exits the adductor canal through the adductor hiatus.
(E) It passes through the obturator foramen.

30. Which of the following arteries supplies most of the blood to the head and neck of the femur?

(A) medial circumflex femoral
(B) obturator
(C) lateral circumflex femoral
(D) external pudendal
(E) deep circumflex iliac

31. All of the following statements concerning the adductor canal are correct EXCEPT

(A) It is approximately 15 cm long.
(B) It extends from the apex of the femoral triangle to the adductor hiatus.
(C) It contains the saphenous nerve.
(D) Is is bounded posteriorly by the sartorius muscle.
(E) It contains the femoral artery and vein.

32. Which of the following statements concerning the lesser sciatic foramen is correct?

(A) It is the passageway for structures entering or leaving the pelvis.
(B) It is the passageway for structures entering or leaving the perineum.
(C) All lower limb arteries and nerves leave the pelvis through this foramen.
(D) The femoral nerve passes through the lesser sciatic foramen.
(E) The obturator nerve enters the adductor compartment via the lesser sciatic foramen.

33. All of the following structures pass through the greater sciatic foramen EXCEPT

(A) piriformis muscle
(B) sciatic nerve
(C) superior gluteal vessels
(D) inferior gluteal vessels
(E) pudendal nerve

34. All of the following statements concerning the gluteus maximus are correct EXCEPT

(A) It is used very little during casual walking.
(B) It assists in making the knee stable.
(C) It is used very little in climbing upstairs.
(D) It is used in running.
(E) It is used when rising from the sitting position.

35. The ischial bursa separates the inferior part of which of the following muscles from the ischial tuberosity?

(A) gluteus maximus
(B) gluteus minimus
(C) gluteus medius
(D) piriformis
(E) obturator internus

36. All of the following statements concerning the gluteus medius and minimus are correct EXCEPT

(A) They all have the same nerve supply.
(B) They have the same actions.
(C) They are supplied by the same blood vessels.
(D) They abduct the thigh and rotate it laterally.
(E) They are largely responsible for preventing sagging of the unsupported side of the pelvis during walking.

37. The positive Trendelenburg sign is associated with injuries to which of the following?

(A) quadriceps femoris
(B) adductor muscles
(C) abductors of the thigh
(D) hamstring muscles
(E) piriformis muscle

38. Which of the following muscles is part of the triceps coxae?

(A) obturator externus
(B) gemelli muscles
(C) piriformis
(D) gluteus medius
(E) quadratus femoris

39. All of the following muscles are lateral rotators of the thigh EXCEPT

(A) quadratus femoris
(B) obturator internus
(C) obturator externus
(D) gluteus medius
(E) inferior gemellus

40. All of the following statements concerning the inferior clunial nerves are correct EXCEPT

(A) They are gluteal branches of the posterior cutaneous nerve of the thigh.
(B) These nerves curl around the inferior border of the gluteus maximus.
(C) They are derivatives of the sacral plexus S1–S3.
(D) They supply the inferior half of the buttock.
(E) They are dorsal primary rami.

41. All of the following nerves are ventral primary rami EXCEPT

(A) posterior cutaneous nerve of the thigh
(B) inferior gluteal
(C) pudendal
(D) sciatic
(E) superior cluneal

42. All of the following statements concerning the sciatic nerve are correct EXCEPT

(A) It is the largest nerve in the body.
(B) It is really two nerves.
(C) It supplies all leg and foot muscles.
(D) It receives its blood supply from the superior gluteal nerve.
(E) It passes inferolaterally under cover of the gluteus maximus, midway between the greater trochanter and the ischial tuberosity.

43. The piriformis syndrome involves which of the following nerves?

(A) sciatic
(B) obturator
(C) femoral
(D) ilioinguinal
(E) inferior gluteal

44. All of the following statements concerning the internal pudendal artery are correct EXCEPT

(A) It does not supply any structures in the gluteal region.
(B) It passes to the perineum with the pudendal nerve.
(C) It supplies the external genitalia.
(D) It crosses the ischial tuberosity.
(E) It re-enters the pelvis through the lesser sciatic foramen.

45. Which of the following muscles is located in the posterior aspect of the thigh?

(A) semitendinosus
(B) gluteus maximus
(C) piriformis
(D) internal obturator
(E) superior gemellus

46. All of the following statements concerning the hamstring muscles are correct EXCEPT

(A) They are extensors of the thigh.
(B) They are flexors of leg.
(C) A person with paralyzed hamstrings tends to fall backwards.
(D) Most take origin from the ischial tuberosity.
(E) Most are innervated by the tibial division of the sciatic nerve.

47. The innervation for the short head of the biceps is provided by which of the following nerves?

(A) obturator
(B) femoral
(C) tibial division of the sciatic
(D) fibular division of the sciatic
(E) sartorius

48. A line drawn from the anterior superior iliac spine to the ischial tuberosity (Nelaton's line), passing over the lateral aspect of the hip, normally passes over which of the following structures?

(A) sciatic nerve
(B) ischial spine
(C) obturator canal
(D) pudendal nerve
(E) greater trochanter

49. All of the following statements concerning the popliteal fossa are correct EXCEPT

(A) The biceps femoris forms the superolateral border.
(B) The semimembranosus forms the superomedial border.
(C) It contains both the tibial and common fibular nerves.
(D) The lateral and medial heads of the gastrocnemius form the inferolateral and inferomedial borders.
(E) It contains the great saphenous vein.

50. The floor of the popliteal fossa includes which of the following structures?

(A) oblique popliteal ligament
(B) patella
(C) lateral meniscus
(D) anterior cruciate ligament
(E) posterior cruciate ligament

51. The lateral sural nerve is a branch of which of the following nerves?

(A) femoral
(B) common fibular
(C) tibial
(D) obturator
(E) posterior cutaneous nerve of the thigh

52. Which of the following muscles dorsiflexes the ankle?

(A) extensor digitorum longus
(B) fibularis longus
(C) soleus

(D) plantaris
(E) gastrocnemius

53. Muscles that evert the foot include which of the following muscles?

(A) gastrocnemius
(B) soleus
(C) tibialis posterior
(D) fibularis brevis
(E) flexor digitorum longus

54. All of the following statements correctly apply to the inferior extensor retinaculum EXCEPT

(A) It passes from the fibula to the tibia proximal to the malleoli.
(B) It is a Y-shaped band of deep fascia.
(C) It attaches laterally to the anterosuperior surface of the calcaneus.
(D) It forms a strong loop around the tendons of the fibularis tertius.
(E) It forms a strong loop around the tendons of the extensor digitorum longus.

55. Which of the following muscles is the strongest dorsiflexor and invertor of the foot?

(A) extensor digitorum longus
(B) tibialis anterior
(C) fibularis tertius
(D) extensor hallucis longus
(E) flexor hallucis longus

56. All of the following muscles are located in the deep muscle group of the posterior compartment EXCEPT

(A) flexor digitorum longus
(B) popliteus
(C) plantaris
(D) flexor hallucis longus
(E) tibialis posterior

57. Which of the following muscles is a flexor of the knee joint?

(A) popliteus
(B) tibialis anterior
(C) fibularis longus
(D) extensor digitorum longus
(E) extensor hallucis longus

58. Which of the following muscles is the powerful "push-off" muscle?

(A) gastrocnemius
(B) soleus
(C) tibialis anterior
(D) flexor hallucis longus
(E) plantaris

59. Which of the following muscles contract to assist the posterior cruciate ligament in preventing anterior displacement of the femur on the tibia?

(A) plantaris
(B) popliteus
(C) soleus
(D) gastrocnemius
(E) tibialis posterior

60. Which of following muscles is included in the triceps surae?

(A) gastrocnemius
(B) tibialis anterior
(C) tibialis posterior
(D) popliteus
(E) fibularis brevis

61. The lateral compartment of the foot contains which of the following muscles?

(A) abductor digiti minimi brevis
(B) flexor hallucis brevis
(C) quadratus plantae
(D) flexor hallucis longus
(E) abductor hallucis

62. The central compartment of the foot contains all of the following muscles EXCEPT

(A) flexor digitorum brevis
(B) flexor digitorum longus
(C) quadratus plantae
(D) abductor hallucis
(E) lumbricals

63. The medial plantar nerve is a terminal branch of which of the following nerves?

(A) femoral
(B) tibial
(C) fibular
(D) obturator
(E) sural

64. All of the following statements concerning the saphenous nerve are correct EXCEPT

(A) It is the largest cutaneous branch of the femoral nerve.
(B) It passes anterior to the medial malleolus to the dorsum of the foot.
(C) It supplies skin along the medial side of the foot.
(D) It supplies skin on the anterior and medial side of the leg.
(E) It innervates the muscles of the foot in the medial compartment of the foot.

65. Which of the following ligaments prevents hyperextension of the hip joint during standing?

(A) iliofemoral
(B) ischiofemoral
(C) pubofemoral
(D) ligament of head of femur
(E) transverse acetabular

66. A synovial protrusion beyond the free margin of the fibrous capsule onto the posterior aspect of the femoral neck forms a bursa for which of these muscle tendons?

(A) piriformis
(B) obturator internus
(C) obturator externus
(D) superior gemellus
(E) quadratus femoris

67. All of the following statements concerning the ligament of the head of the femur are correct EXCEPT

(A) It is a weak ligament.
(B) It is of little importance in strengthening the hip.
(C) It contains a small nerve.
(D) Usually it contains a small artery.
(E) Its wide end attaches to the margins of the acetabular notch and the transverse ligament.

68. Which of the following arteries provides the main blood supply for the hip joint?

(A) obturator
(B) medial circumflex
(C) lateral circumflex
(D) common iliac
(E) internal iliac

69. Which of the following muscles is the most important muscle in stabilizing the knee?

(A) biceps femoris
(B) adductor magnus
(C) obturator internus
(D) quadriceps femoris
(E) piriformis

70. Which of the following muscles passes out of the knee joint to reach the tibia?

(A) plantaris
(B) popliteus
(C) sartorius
(D) gracilis
(E) tibialis anterior

71. All of the following statements concerning the fibular collateral ligament are correct EXCEPT

(A) It splits the tendon of the biceps femoris.
(B) It is superficial to the tendon of the popliteus.
(C) It is connected to the lateral meniscus.
(D) It is rounded and cordlike.
(E) It extends from the lateral epicondyle of the femur to the head of the fibula.

72. Which of the following statements concerning the tibial collateral ligament is correct?

(A) It is attached to the lateral meniscus.
(B) It is stronger than the fibular collateral ligament.

(C) It is less frequently damaged than the fibular collateral ligament.
(D) The tibial collateral ligament and medial meniscus are commonly torn during contact sports such as football.
(E) It is an expansion of the tendon of the adductor magnus.

73. The oblique popliteal ligament is an expansion of the tendon of which of the following muscles?

(A) popliteus
(B) plantaris
(C) adductor magnus
(D) semimembranosus
(E) gastrocnemius

74. All of the following ligaments of the knee are intra-articular EXCEPT

(A) anterior cruciate
(B) lateral meniscus
(C) patellar
(D) posterior cruciate
(E) medial meniscus

75. Which of the following statements concerning the anterior cruciate ligament is correct?

(A) It is the stronger of the two cruciate ligaments.
(B) It has a relatively poor blood supply.
(C) It is the main stabilizing factor for the femur when one is walking downhill.
(D) It tightens during flexion of the knee joint, preventing anterior displacement of the tibia on the femur.
(E) It attaches to the anterior part of the lateral surface of the medial condyle of the femur.

76. All of the following statements concerning the menisci of the knee joint are correct EXCEPT

(A) They are thicker at their external margins.
(B) They taper to thin, unattached edges in the interior of the joint.
(C) They are wedge-shaped in transverse section.
(D) Their external margins attach to the fibrous capsule of the knee joint.
(E) The transverse ligament of the knee joins the posterior edges of the menisci.

77. Which of the following statements correctly applies to the lateral meniscus?

(A) It is larger and less movable than the medial meniscus.
(B) It is in contact with the fibular collateral ligament.
(C) It is attached to the posterior cruciate ligament by the posterior meniscofemoral ligament.
(D) It adheres to the deep surface of the tibial collateral ligament.
(E) It acts like a shock absorber.

78. Which of the following genicular branches supplies the cruciate ligaments?

(A) femoral
(B) popliteal
(C) anterior recurrent branches of the anterior tibial recurrent
(D) circumflex fibular
(E) posterior recurrent branches of the anterior tibial recurrent

79. Which of the following three ligaments are damaged in the "unhappy triad" of knee injuries?

(A) ACL, tibial collateral, and medial meniscus
(B) PCL, fibular collateral, and lateral meniscus
(C) ACL, fibular collateral, and medial meniscus
(D) PCL, tibial collateral, and lateral meniscus
(E) ACL, tibial collateral, and lateral meniscus

80. Pain on lateral rotation of the tibia on the femur indicates damage to which of the following structures?

(A) anterior cruciate ligament
(B) posterior cruciate ligament
(C) lateral meniscus
(D) medial meniscus
(E) posterior meniscofemoral ligament

81. All of the following ligaments reinforce the fibrous capsule on the medial side of the ankle EXCEPT

(A) tibiocalcaneal
(B) posterior tibiotalar
(C) anterior tibiotalar
(D) tibionavicular
(E) calcaneofibular

82. Which of the following groups of muscles produce dorsiflexion of the ankle?

(A) anterior compartment of the leg
(B) posterior compartment of the leg
(C) medial compartment of the foot
(D) lateral compartment of the leg
(E) lateral compartment of the foot

83. All of the following statements concerning the plantar calcaneonavicular ligament are correct EXCEPT

(A) It extends from the sustentaculum tali to the navicular.
(B) It is also known as the spring ligament.
(C) It lies deep to the plantar aponeurosis.
(D) It plays an important role in maintaining the longitudinal arch of the foot.
(E) It forms a tunnel for the tendon of the fibularis longus.

84. All of the following statements concerning the medial longitudinal arch of the foot are correct EXCEPT

(A) It is higher and more important than the lateral longitudinal arch.
(B) It is composed of the calcaneus, talus, navicular, cuneiforms, and three metatarsals.
(C) The calcaneus is the keystone of this arch.
(D) The tibialis anterior helps strengthen this arch.
(E) The fibularis longus tendon also supports this arch.

85. Which of the following conditions is associated with lateral deviation of the great toe?

(A) hallux valgus
(B) hammer toe
(C) claw toe
(D) pes planus
(E) club foot

86. All of the following statements concerning the transverse tarsal joint are correct EXCEPT

(A) It occurs where the talus rests on and articulates with the calcaneus.
(B) It is formed by the combined talonavicular part of the talocalcaneonavicular and calcaneocuboid joints.
(C) Transection across this joint is a standard method for surgical amputation of the foot.
(D) It is two separate joints aligned transversely.
(E) Dorsiflexion and plantarflexion of the foot are the main movements involving this joint.

87. Which of the of the following areas is involved in tibial nerve entrapment?

(A) medial malleolus to the calcaneus
(B) lateral malleolus to the navicular bone
(C) medial malleolus to the calcaneal tendon
(D) lateral malleolus to the fibularis brevis
(E) long plantar ligament to the tendon of the fibularis longus

88. Which of the following major joints is the most frequently injured?

(A) shoulder
(B) elbow
(C) hip
(D) knee
(E) ankle

89. Which of the following statements correctly applies in Pott's fracture-dislocation?

(A) The foot is forcibly dorsiflexed.
(B) The foot is forcibly plantar flexed.
(C) The foot is forcibly inverted.
(D) The foot is forcibly everted.
(E) The calcaneus is usually fractured.

90. The grip of the malleoli on the trochlea is strongest during which of the following movements of the ankle?

(A) plantarflexion
(B) dorsiflexion
(C) eversion
(D) inversion
(E) lateral rotation

91. Which of the following statements correctly applies to genu valgum?

(A) The tibia is diverted medially.
(B) The tibia is diverted laterally.
(C) The medial side of the knee takes all the pressure.
(D) This deformity causes wear and tear of the medial meniscus.
(E) This deformity does not influence weight distribution.

92. Which of the following knee support structures is considered to be the most important in the stabilization of the joint?

(A) lateral and medial menisci
(B) anterior and posterior cruciate ligaments
(C) medial and lateral collateral ligaments
(D) fibrous capsule
(E) quadriceps femoris

93. Which of the following fractures is the most troublesome and problematic?

(A) medial malleolus of tibia
(B) lateral malleolus of fibula
(C) sustentaculum of calcaneus
(D) femoral neck
(E) adductor tubercle

94. Which of the following muscles is the strongest flexor of the hip joint?

(A) semitendinosus
(B) iliopsoas
(C) gluteus medius
(D) gracilis
(E) pectineus

95. All of the following muscles are lateral rotators of the hip joint EXCEPT

(A) obturator externus
(B) superior gemellus
(C) piriformis
(D) gluteus minimus
(E) quadratus femoris

96. Which of the following arteries is usually evaluated during a physical examination of the peripheral vascular system?

(A) lateral plantar
(B) dorsalis pedis
(C) popliteal
(D) fibular
(E) posterior tibial

97. The tendon of the biceps femoris and the neck of the fibula may be used as a guide for locating which of the following nerves?

(A) saphenous
(B) sural
(C) common fibular
(D) medial plantar
(E) tibial

98. Which of the following muscle tendons is commonly removed for grafting without causing disability?

(A) extensor hallucis brevis
(B) fibularis tertius
(C) plantaris
(D) flexor digitorum brevis
(E) tendons of the flexor digitorum longus

99. Shin splints usually involve muscles in which of the following compartments?

(A) anterior
(B) lateral
(C) posterior
(D) medial plantar
(E) lateral plantar

100. The strongest dorsiflexor of the foot is which of the following muscles?

(A) fibularis tertius
(B) extensor digitorum longus
(C) tibialis anterior
(D) gastrocnemius
(E) extensor hallucis brevis

DIRECTIONS (Questions 101 through 105): Identify the anatomical features indicated on the art below.

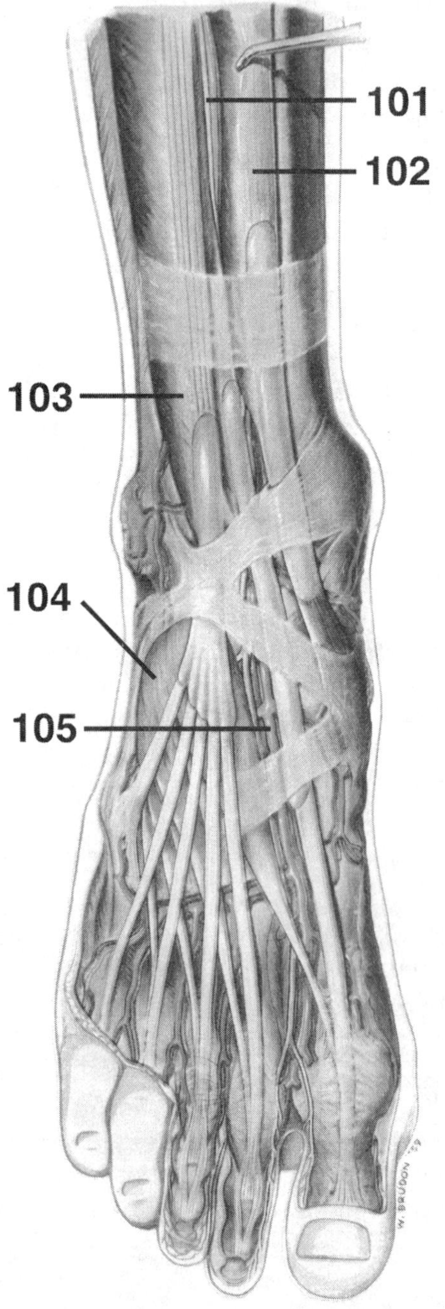

DIRECTIONS (Questions 106 through 110): Identify the anatomical features indicated on the art below.

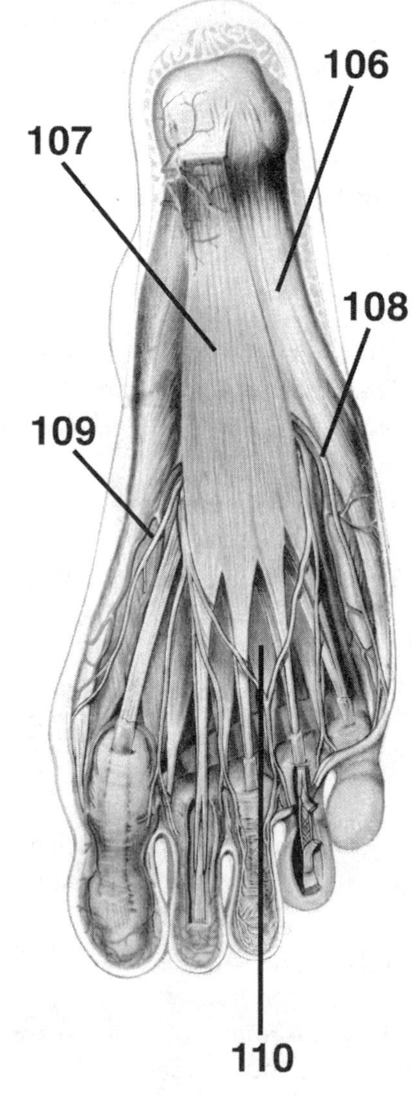

DIRECTIONS (Questions 111 through 115): Identify the anatomical features indicated on the art below.

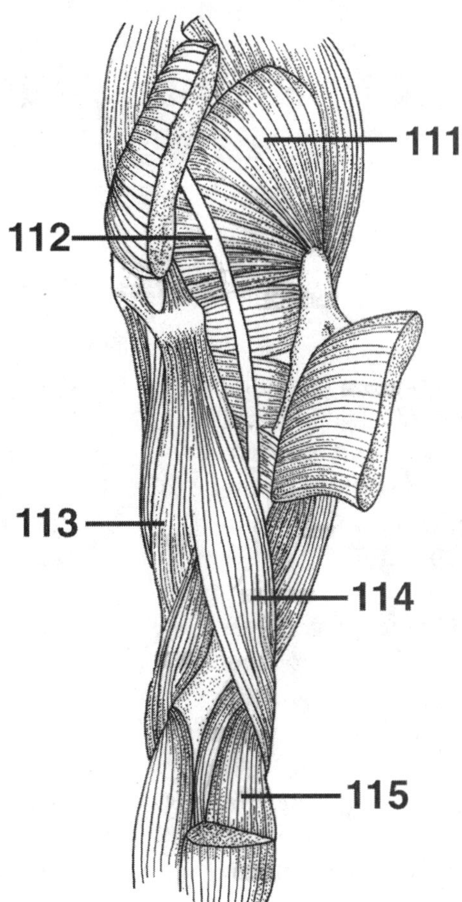

Answers and Explanations

1. **(E)** The femur is the longest and heaviest bone in the body. In the average adult, its length is approximately one-quarter of the person's height (108 cm or 18 in.). The distal end of the femur undergoes ossification just before birth *(Moore, p 509).*

2. **(A)** The medial surface of the medial condyle has a large and prominent medial epicondyle, superior to which is another elevation, the adductor tubercle *(Moore, p 509).*

3. **(D)** The margin of the acetabulum is deficient inferiorly at the acetabular notch, which makes the fossa resemble a cup with a piece of its lip missing *(Moore, p 508).*

4. **(C)** The acetabular notch and fossa comprise a deficiency in the smooth articular lunate surface of the acetabulum, which actually articulates with the head of the femur *(Moore, p 508).*

5. **(D)** A fracture of the femoral neck is among the most troublesome and problematic of all fractures because of the instability of the fracture site. Furthermore, the periosteum covering the femoral neck is exceedingly thin and has extremely limited powers of osteogenesis (bone formation). Because the retinacular arteries arise from the medial circumflex femoral arteries and run parallel to the femoral neck on their way to supply the femoral head, they are vulnerable to injury when the neck of the femur fractures. Rupture of these vessels cause degeneration (necrosis) of the femoral head and bleeding into the hip joint *(Moore, p 511).*

6. **(B)** Elderly people fracture the neck of the femur and refer to the injury as a "broken hip" *(Moore, p 511).*

7. **(C)** The lateral surface of the medial malleolus articulates with the talus, and the lateral malleolus helps hold the talus in its socket *(Moore, p 513).*

8. **(B)** The body of the tibia is the most common site for a compound fracture—one in which the skin is perforated and blood vessels are torn *(Moore, p 513).*

9. **(B)** The calcaneus is the largest and strongest bone in the foot. The calcaneus transmits most of the body weight from the talus to the ground *(Moore, p 515).*

10. **(E)** The shelf-like sustentaculum tali projects from the superior border of the medial surface of the calcaneus and supports the talar head *(Moore, p 515).*

11. **(A)** The talus articulates with the fibula, calcaneus, navicular, and tibia *(Moore, p 515).*

12. **(C)** The 1st metatarsal is shorter and stouter than the others. The 2nd metatarsal is the longest. The bases of the metatarsals articulate with the cuneiform and cuboid bones. The base of the 5th metatarsal has a large tuberosity *(Moore, p 515).*

13. **(D)** The deep fascia of the thigh is called fascia lata and the deep fascia of the leg is called

crural fascia. Scarpa's fascia is the membranous fascia of the lower abdominal wall and Colles' fascia is located in the perineum *(Moore, p 522).*

14. **(C)** The saphenous opening in the fascia lata is a deficiency in the deep fascia lata inferior to the medial part of the inguinal ligament, approximately 4 cm inferolateral to the pubic tubercle. The medial margin of the saphenous opening is smooth, but its superior, lateral, and inferior margins form a sharp crescentic edge, the falciform margin. The sickle-shaped margin of the saphenous opening is joined at its medial margin by fibrofatty tissue—the cribriform fascia *(Moore, p 524).*

15. **(E)** The great saphenous vein passes through the saphenous opening and cribriform fascia to enter the femoral vein. Some efferent lymphatic vessels from the superficial inguinal lymph nodes also pass through the saphenous opening and cribriform fascia to enter the deep inguinal lymph nodes *(Moore, p 524).*

16. **(D)** The small saphenous vein ascends posterior to the lateral malleolus and passes along the lateral border of the calcaneal tendon. It then ascends between the heads of the gastrocnemius muscle and empties into the popliteal vein in the popliteal fossa *(Moore, p 526).*

17. **(B)** The iliopsoas is the chief flexor of the thigh, and when the thigh is fixed, it flexes the trunk on the hip. Its broad lateral part, the iliacus, and its long medial part, the psoas major, arise from the iliac fossa and lumbar vertebrae, respectively. It is an anterior thigh muscle *(Moore, p 531).*

18. **(D)** The sartorius is known as the "tailor's muscle." It is the longest muscle in the body and acts across two joints. It flexes the hip joint and participates in flexion of the knee. It is located in the anterior compartment of the thigh *(Moore, p 531).*

19. **(B)** The quadriceps femoris forms the main bulk of the anterior thigh muscles and collectively constitutes the largest and one of the most powerful muscles in the body. The quadriceps is the great extensor of the leg, and all four of its parts combine to from a tendinous attachment to the tibia. The patella also provides additional leverage for the quadriceps *(Moore, pp 532–534).*

20. **(A)** The rectus femoris is considered to be the "kicking muscle." The vastus lateralis is the largest component of the quadriceps. The vastus intermedius lies deep to the rectus femoris, and the articular muscle is considered to be a derivative of the vastus intermedius *(Moore, p 534).*

21. **(B)** The long, strap-like muscle lies along the medial side of the thigh and knee. It is the only muscle of the adductor group that crosses the knee. It is the most superficial of the adductor group and is the weakest member *(Moore, p 538).*

22. **(B)** The adductor magnus is the largest muscle in the adductor group. It is located in the medial compartment of the thigh. This large adductor is a composite, triangular muscle that has adductor and hamstring parts. The two parts differ in their attachments, nerve supply, and main actions *(Moore, p 538).*

23. **(D)** The adductor hiatus is an opening in the aponeurotic distal attachment of the adductor magnus. It transmits the femoral artery and vein from the adductor canal in the thigh to the popliteal fossa posterior to the knee. The opening is just superior to the adductor tubercle of the femur *(Moore, p 541).*

24. **(E)** The femoral triangle is bounded superiorly by the inguinal ligament, medially by the adductor longus, and laterally by the sartorius. The femoral triangle is bisected by the femoral artery and vein, which leave and enter the adductor canal at its apex. The saphenous nerve descends through the femoral triangle *(Moore, p 541).*

25. **(C)** The femoral sheath extends 3 to 4 cm inferior to the inguinal ligament and encloses proximal parts of the femoral vessel and the

femoral canal. The sheath is formed by an inferior prolongation of transversalis and iliopsoas fascia. The femoral sheath does not enclose the femoral nerve. The sheath ends by becoming continuous with the adventitia of the femoral vessels *(Moore, p 542).*

26. **(A)** The medial compartment of the femoral sheath is the femoral canal. It extends distally to the level of the proximal edge of the saphenous opening. It allows the femoral vein to expand when venous return from the lower limb is increased. It contains loose connective tissue, fat, a few lymphatic vessels, and sometimes a deep inguinal lymph node (Cloquet's node) *(Moore, pp 542–543).*

27. **(C)** The femoral ring is closed at the proximal end by extraperitoneal fatty tissue, which forms the femoral septum. The boundaries of the femoral ring include the partition between the femoral canal and the femoral vein laterally. Posteriorly, the superior ramus of the pubis is covered by the pectineus muscle. Its medial boundary is the lacunar ligament; anteriorly, the boundary is the medial part of the inguinal ligament *(Moore, pp 543–545).*

28. **(B)** The femoral artery is the chief artery of the lower limb and is the continuation of the external iliac artery. It bisects the femoral triangle at its apex and enters the adductor canal deep to the sartorius muscle *(Moore, p 545).*

29. **(A)** The deep artery of the thigh is the largest branch of the femoral artery and the chief artery of the thigh. It arises in the femoral triangle from the lateral side of the femoral artery *(Moore, p 545).*

30. **(A)** The medial circumflex femoral artery is especially important because it supplies most of the blood to the head and neck of the femur *(Moore, p 545).*

31. **(D)** The adductor canal (Hunter's canal) is approximately 15 cm long and is a narrow fascial tunnel in the thigh running from the apex of the femoral triangle to the adductor hiatus in the tendon of the adductor magnus. Located deep or posterior to the middle third of the sartorius, the adductor canal provides an intermuscular passage through which the femoral vessels pass to reach the popliteal fossa. The contents of the adductor canal include the femoral vessels, saphenous nerve, and nerve to the vastus medialis *(Moore, p 549).*

32. **(B)** The greater sciatic foramen is the passageway for structures entering or leaving the pelvis, whereas the lesser sciatic foramen is the passageway for structures entering or leaving the perineum. The greater sciatic foramen is the opening for all lower limb arteries and nerves leaving the pelvis and entering the gluteal region *(Moore, p 550).*

33. **(E)** The greater sciatic foramen is the passageway for the sciatic nerve, piriformis muscle, and gluteal vessels. The pudendal nerve enters the perineum through the lesser sciatic foramen *(Moore, p 553).*

34. **(C)** The gluteus maximus is used when rising from the sitting position or straightening from the bending position. It is used in climbing steps and running. It also assists in making the knee stable. It is used very little during casual walking and when one is standing motionless *(Moore, p 552).*

35. **(A)** The ischial bursa separates the inferior part of the gluteus maximus from the ischial tuberosity, which is often absent *(Moore, p 552).*

36. **(D)** The gluteus medius and minimus have the same actions and nerve supply and are supplied by the superior gluteal arteries. Both muscles abduct the thigh and rotate it medially. They play an essential role during locomotion and are largely responsible for preventing sagging of the unsupported side of the pelvis during walking *(Moore, p 554).*

37. **(C)** When the weight is on both feet, the pelvis is evenly supported and does not sag. When the weight is borne by one foot, the muscles on the same side hold the pelvis so the pelvis will not sag on the side of the raised foot. When the gluteus medius and minimus (abductors of the

thigh) are inactive owing to injury of the superior gluteal nerve, the supporting and steadying action of these muscles is lost and the pelvis falls on the side of the raised limb. This is referred to as a positive Trendelenburg sign *(Moore, p 555).*

38. **(B)** The obturator internus and the superior and inferior gemelli form a tricipital (three-headed) muscle that is sometimes called the triceps coxae *(Moore, p 550).*

39. **(D)** The gluteus maximus, piriformis, obturator internus and externus, and superior and inferior gemelli are all lateral rotators of the thigh. Both the gluteus medius and the gluteus minimus are medial rotators of the thigh *(Moore, p 551).*

40. **(E)** The inferior clunial nerves are gluteal branches of the posterior cutaneous nerve of the thigh, a derivative of the sacral plexus (ventral rami S1 through S3). These nerves curl around the inferior border of the gluteus maximus and supply the inferior half of the buttock *(Moore, p 556).*

41. **(E)** The superior cluneal nerves are dorsal rami of L1–L3 and the middle cluneal nerves are dorsal rami of S1–S3. The sciatic, pudendal, posterior cutaneous nerve of the thigh, and the gluteal nerves are all ventral primary rami *(Moore, p 557).*

42. **(D)** The sciatic nerve is the largest nerve in the body. It is so large that it receives its own blood supply from the inferior gluteal artery. It runs inferolaterally under cover of the gluteus maximus, midway between the greater trochanter and the ischial tuberosity. The sciatic nerve is really two nerves, the tibial and common fibular *(Moore, p 558).*

43. **(A)** A pain in the buttock may result from compression of the sciatic nerve by the piriformis muscle (piriformis syndrome) *(Moore, p 559).*

44. **(D)** The internal pudendal artery leaves the gluteal region immediately by crossing the ischial spine and re-entering the pelvis through the lesser sciatic foramen. The artery passes to the perineum with the pudendal nerve and supplies the external genitalia and muscles in the pelvic region. It does not supply any structures in the gluteal region *(Moore, p 562).*

45. **(A)** The three muscles in the posterior aspect of the thigh are the hamstrings, which include the semitendinosus, semimembranosus, and biceps femoris *(Moore, p 563).*

46. **(C)** The hamstring muscles arise from the ischial tuberosity and are innervated by the tibial division of the sciatic nerve. The short head of the biceps does not meet these criteria. The hamstrings are extensors of the thigh and flexors of the leg. A person with paralyzed hamstrings tends to fall forward because the gluteus maximus muscles cannot maintain the necessary muscle tone to stand straight *(Moore, p 563).*

47. **(D)** The long head of the biceps femoris is innervated by the tibial division of the sciatic nerve and the short head of the biceps is innervated by the fibular division of the sciatic nerve *(Moore, p 563).*

48. **(E)** A line drawn from the anterior superior iliac spine to the ischial tuberosity (Nélaton's line), passing over the lateral aspect of the hip region, normally passes over or near the top of the greater trochanter *(Moore, p 567).*

49. **(E)** The biceps femoris forms the superolateral border and the semimembranosus muscle forms the superomedial border. The medial and lateral heads of the gastrocnemius form the inferolateral and inferomedial borders. The contents of the fossa include the small saphenous vein, popliteal arteries and veins, and tibial and common fibular nerves *(Moore, p 571).*

50. **(A)** The floor of the popliteal fossa is formed by the popliteal surface of the femur, the oblique popliteal ligament, and the popliteal fascia over the popliteus *(Moore, p 571).*

51. **(B)** The medial sural nerve is derived from the tibial nerve and the lateral sural nerve is

derived from the common fibular. The medial and lateral sural nerves unite to form the sural nerve *(Moore, p 590).*

52. **(A)** Dorsiflexors of the ankle include the tibialis anterior, extensors digitorum longus, hallucis longus, and fibularis tertius *(Moore, p 577).*

53. **(D)** Both the fibularis longus and brevis evert the foot. The fibularis tertius also aids in eversion of the foot *(Moore, p 577).*

54. **(A)** The superior extensor retinaculum is a strong, broad band of deep fascia passing from the fibula to the tibia, proximal to the malleoli. The inferior extensor retinaculum is a Y-shaped band of deep fascia that attaches laterally to the anterosuperior surface of the calcaneus. It forms a strong loop around the tendons of the fibularis tertius and the extensor digitorum longus muscle *(Moore, p 579).*

55. **(B)** The tibialis anterior is the strongest dorsiflexor and inverter of the foot *(Moore, p 579).*

56. **(C)** The plantaris is located in the superficial group of muscles in the posterior compartment *(Moore, p 586).*

57. **(A)** The popliteus is a flexor of the knee joint *(Moore, p 588).*

58. **(D)** The flexor hallucis longus is the powerful "push-off" muscle during walking, running, and jumping *(Moore, p 589).*

59. **(B)** When a person is standing with the knee partly flexed, the popliteus contracts to assist the posterior cruciate ligament in preventing anterior displacement of the femur on the tibia *(Moore, p 588).*

60. **(A)** Together, the two-headed gastrocnemius and soleus form the three-headed triceps surae *(Moore, p 586).*

61. **(A)** The lateral compartment of the foot contains the abductor and flexor digiti minimi brevis *(Moore, p 596).*

62. **(D)** The central compartment of the foot contains the flexor digitorum brevis, flexor digitorum longus, quadratus plantae, lumbricals, proximal part of the tendon flexor hallucis longus, and the lateral plantar nerve and vessel *(Moore, p 596).*

63. **(B)** The medial plantar nerve, the larger of the two terminal branches of the tibial nerve, passes deep to the abductor hallucis and runs anteriorly between the muscles and the flexor digitorum brevis on the lateral side of the medial plantar artery *(Moore, p 601).*

64. **(E)** The saphenous nerve is the largest cutaneous branch of the femoral nerve. In addition to supplying the skin and fascia on the anterior and medial sides of the leg, the saphenous nerve passes anterior to the medial malleolus to the dorsum of the foot, where it supplies skin along the medial side of the foot as far as the head of the 1st metatarsal *(Moore, p 601).*

65. **(A)** The fibrous capsule is reinforced anteriorly by the strong, Y-shaped iliofemoral ligament (of Bigelow) which attaches to the anterior inferior iliac spine and the acetabular rim proximally and the intertrochanteric line distally. The iliofemoral ligament prevents hyperextension of the hip during standing by screwing the femoral head into the acetabulum *(Moore, p 611).*

66. **(C)** A synovial protrusion beyond the free margin of the fibrous capsule onto the posterior aspect of the femoral neck forms a bursa for the obturator externus tendon *(Moore, p 611).*

67. **(C)** The ligament of the head of the femur is weak and of little importance in strengthening the hip joint. Its wide end attaches to the margins of the acetabular notch and the transverse acetabular ligament. Its narrow end attaches to the pit in the head of the femur. Usually the ligament contains a small artery to the head of the femur *(Moore, p 611).*

68. **(B)** The main blood supply of the hip joint is from branches of the circumflex femoral arteries (especially the medial circumflex femoral artery) that travel in the retinacula (reflections

of the capsule along the neck of the femur toward the head). These retinacular vessels may be damaged in femoral neck fractures and result in avascular necrosis of the femoral head *(Moore, p 613).*

69. **(D)** The most important muscle in stabilizing the knee joint is the large quadriceps femoris, particularly inferior fibers of the vastus medialis and lateralis. The knee joint will function surprisingly well following a ligament strain if the quadriceps is well conditioned *(Moore, pp 617–618).*

70. **(B)** The fibrous capsule of the knee is deficient on the lateral condyle to allow the tendon of the popliteus to pass out of the joint to attach to the tibia *(Moore, p 618).*

71. **(C)** The fibular collateral ligament (lateral collateral ligament), rounded and cordlike, is strong. It extends inferiorly from the lateral epicondyle of the femur to the lateral surface of the head of fibula. The tendon of the popliteus passes deep to the fibular collateral ligament, separating it from the lateral meniscus. The tendon of the biceps femoris is also split into two parts by this ligament *(Moore, p 619).*

72. **(D)** The tibial collateral ligament (medial collateral ligament) is a strong, flat band that extends from the medial epicondyle of the femur to the medial surface of the tibia. At its midpoint, the deep fibers of the tibial collateral ligament are firmly attached to the medial meniscus. The tibial collateral ligament, weaker than the fibular collateral ligament, is more often damaged. As a result, the tibial collateral ligament and medial meniscus are commonly torn during contact sports such as football *(Moore, p 619).*

73. **(D)** The oblique popliteal ligament is an expansion of the tendon of the semimembranosus, which strengthens the fibrous capsule posteriorly *(Moore, p 619).*

74. **(C)** The patellar ligament is extracapsular. The cruciate ligaments and menisci are classified as intra-articular and are found within the knee joint. The popliteal tendon is also intra-articular during part of its course *(Moore, p 620).*

75. **(B)** The anterior cruciate ligament (ACL), the weaker of the two cruciate ligaments, arises from the anterior intercondylar area of the tibia, just posterior to the attachment of the medial meniscus. It extends superiorly, posteriorly, and laterally to attach to the posterior part of the medial side of the lateral condyle of the femur. The ACL has a relatively poor blood supply. It is slack when the knee is flexed and taut when it is fully extended, preventing posterior displacement. *(Moore, p 620).*

76. **(E)** The menisci are thicker at their external margins and taper to thin, unattached edges in the interior of the joint. Wedge-shaped in transverse section, the menisci are firmly attached at their ends to the intercondylar area of the tibia. Their external margins attach to the fibrous capsule of the knee joint. The transverse ligament of the knee, a slender, fibrous band, joins the anterior edges of the menisci, allowing them to move together during knee movements *(Moore, p 621).*

77. **(D)** The lateral meniscus is nearly circular and is smaller and more movable than the medial meniscus. The tendon of the popliteus separates the lateral meniscus from the fibular collateral ligament. A strong tendinous slip, the posterior meniscofemoral ligament, joins the lateral meniscus to the posterior cruciate ligament and the medial femoral condyle. The lateral meniscus acts like a shock absorber *(Moore, p 621).*

78. **(B)** The middle genicular branches of the popliteal artery penetrate the fibrous capsule of the knee joint and supply the cruciate ligaments, synovial membrane, and peripheral margins of the menisci *(Moore, p 626).*

79. **(A)** The ACL may tear when the tibial collateral ligament ruptures. First, the tibial collateral ligament ruptures, opening the joint on the medial side and possibly tearing the medial meniscus and ACL. This "unhappy triad of injuries" can result from clipping in football *(Moore, p 626).*

80. **(C)** Pain on lateral rotation of the tibia on the femur indicates injury of the lateral meniscus, whereas pain on medial rotation of the tibia on the femur indicates injury of the medial meniscus *(Moore, p 628).*

81. **(E)** The lateral ligament consists of the anterior and posterior talofibular ligaments and the calcaneofibular ligament. The three discrete ligaments are collectively referred to as the lateral ligament. The fibrous capsule is reinforced medially by the large, strong medial ligament (deltoid ligament), which attaches proximally to the medial malleolus. This ligament consists of the tibionavicular, tibiocalcaneal, and anterior and posterior tibiotalar ligaments *(Moore, pp 633–635).*

82. **(A)** Dorsiflexion of the ankle is produced by the muscles in the anterior compartment of the leg *(Moore, p 635).*

83. **(E)** The plantar calcaneonavicular ligamentum (spring ligament) extends from the sustentaculum tali to the posteroinferior surface of the navicular. It plays an important role in maintaining the longitudinal arch of the foot *(Moore, p 637).*

84. **(C)** The talar head is the keystone of the medial longitudinal arch *(Moore, p 640).*

85. **(A)** Hallux valgus is a foot deformity characterized by lateral deviation of the great toe. Hammertoe is a deformity in which the proximal phalanx is permanently flexed at the metatarsophalangeal joint and the middle phalanx is plantarflexed at the interphalangeal joint. The distal phalanx is also flexed. Claw toes are characterized by hyperextension of the metatarsophalangeal joints and flexion of the distal interphalangeal joints. Pes planus is flat feet, and *clubfoot* refers to a foot that is twisted. The foot is inverted, the ankle is plantarflexed, and the forefoot is adducted *(Moore, pp 641–642).*

86. **(E)** The transverse tarsal joint is formed by the combined talonavicular part of the talocalcaneonavicular and calcaneocuboid joints, two separate joints aligned transversely. Transection across the transverse tarsal joint is a standard method for surgical amputation of the foot. Inversion and eversion of the foot are the main movements involving these joints *(Moore, p 637).*

87. **(A)** The tibial nerve leaves the posterior compartment of the leg by passing deep to the flexor retinaculum in the interval between the medial malleolus and calcaneus. The area involved is from the medial malleolus to the calcaneus, and the heel pain results from compression of the tibial nerve by the flexor retinaculum *(Moore, p 636).*

88. **(E)** The ankle is the most frequently injured major joint in the body. Ankle sprains are most common. A sprained ankle is nearly always an inversion injury *(Moore, p 636).*

89. **(D)** A Pott's fracture-dislocation of the ankle occurs when the foot is forcibly everted. This action pulls on the extremely strong medial ligament, often tearing off the medial malleolus. The talus then moves laterally, shearing off the lateral malleolus or, more commonly, breaking the fibula superior to the inferior tibiofibular joint *(Moore, p 636).*

90. **(B)** The grip of the malleoli on the trochlea is strongest during dorsiflexion of the foot, because this movement forces the wider, anterior part of the trochlea posteriorly, spreading the tibia and fibula slightly apart *(Moore, p 632).*

91. **(B)** Genu varum and genu valgum result in deviation of the tibia from the midline. In genu varum, the tibia is diverted medially, and in genu valgum, the tibia is diverted laterally; these deformities cause unequal weight distribution. In the varum deformity, the medial side of the knee takes all the pressure, leading to wear and tear of the medial meniscus *(Moore, p 630).*

92. **(E)** The stability of the knee joint depends on the strength and actions of the surrounding muscles and their tendons. The ligaments that connect the femur and tibia are also important. Of these supports, the muscles are most impor-

tant; therefore, many sport injuries are preventable through appropriate conditioning and training. The most important muscle in stabilizing the knee joint is the large quadriceps femoris, particularly inferior fibers of the vastus medialis and lateralis. The knee joint will function surprisingly well following a ligament strain if the quadriceps is well conditioned *(Moore, pp 617–618).*

93. **(D)** Fractures of the femoral neck are intracapsular, and realignment of the neck fragments requires internal skeletal fixation. Femoral neck fractures are among the most troublesome and problematic of all fractures *(Moore, p 614).*

94. **(B)** The iliopsoas is the strongest flexor of the hip joint *(Moore, p 613).*

95. **(D)** Medial rotators of the hip joint include anterior fibers of the gluteus medius, gluteus minimus, and tensor fascia lata; lateral rotators include the obturator externus, obturator internus, superior gemellus, piriformis, quadratus femoris, and gluteus maximus *(Moore, p 613).*

96. **(B)** The pulse of the dorsalis pedis artery, or dorsal artery of the foot, is evaluated during a physical examination of the peripheral vascular system. Dorsalis pedis pulses may be palpated with the feet slightly dorsiflexed *(Moore, pp 603–604).*

97. **(C)** The tendon of the biceps femoris may be traced by palpating its distal attachment to the lateral side of the head of the fibula. This tendon and the neck of the fibula guide the examining finger to the common fibular nerve *(Moore, p 592).*

98. **(C)** Because of its minor role, the plantaris tendon can be removed for grafting without causing any disability *(Moore, p 586).*

99. **(A)** Shin splints commonly result from traumatic injury or athletic overexertion of muscles in the anterior compartment—especially the tibialis anterior *(Moore, p 580).*

100. **(C)** The tibialis anterior is the strongest dorsiflexor and invertor of the foot *(Moore, p 579).*

101. deep peroneal nerve

102. tibialis anterior

103. extensor digitorum longus

104. extensor digitorum brevis

105. dorsalis pedis artery

106. abductor digiti minimi

107. flexor digitorum brevis

108. lateral plantar nerve

109. medial plantar nerve

110. lumbrical

111. gluteus minimus

112. sciatic nerve

113. semitendinosus

114. long head of biceps femoris

115. soleus

CHAPTER 7

The Head and Neck

Questions

DIRECTIONS (Questions 1 through 146): Each of the numbered items or incomplete statements in this section is followed by answers or by completions of the statement. Select the ONE lettered answer or completion that is BEST in each case.

1. Which of the following is NOT a bone of the neurocranium?

(A) palatine bones
(B) parietal bones
(C) sphenoid bones
(D) temporal bones
(E) ethmoid bone

2. Which of the following is NOT a bone of the facial skeleton?

(A) maxilla
(B) zygomatic bone
(C) frontal bone
(D) inferior nasal concha
(E) nasal bone

3. The metopic suture is a persistence of which of the following?

(A) frontal suture
(B) coronal suture
(C) sagittal suture
(D) lambdoid suture
(E) hypophyseal suture

4. The external occipital protuberance is also known as which of the following?

(A) nasion
(B) inion
(C) bregma
(D) pterion
(E) vertebra prominens

5. Which of the following best describes the landmark known as lambda?

(A) point on calvaria at junction of sagittal and lambdoid sutures
(B) point on calvaria at junction of sagittal and coronal sutures
(C) junction of the greater wing of the sphenoid, squamous temporal, frontal, and parietal bones
(D) star-shaped landmark at junction of parietomastoid, occipitomastoid, and lambdoid sutures
(E) smooth prominence on frontal bone superior to root of nose

6. The superior point of the neurocranium in the midline is known as which of the following?

(A) pterion
(B) bregma
(C) vertex
(D) asterion
(E) nasion

7. Which of the following foramina is NOT in the middle cranial fossa?

(A) foramen rotundum
(B) foramen spinosum
(C) foramen lacerum
(D) groove of greater petrosal nerve
(E) foramen magnum

8. Which of the following foramina is located in the anterior cranial fossa?
   (A) foramen cecum
   (B) optic canals
   (C) superior orbital fissures
   (D) foramen ovale
   (E) condylar canal

9. Which of the following foramina does NOT transmit emissary veins?
   (A) foramen cecum
   (B) condylar canal
   (C) mastoid foramen
   (D) parietal foramen
   (E) anterior ethmoidal foramina

10. The superior orbital fissure transmits all of the following EXCEPT
    (A) ophthalmic division of the trigeminal nerve
    (B) maxillary division of the trigeminal nerve
    (C) oculomotor nerve
    (D) trochlear nerve
    (E) abducens nerve

11. All of the following transmit an arterial branch to the meninges EXCEPT
    (A) foramen ovale
    (B) foramen rotundum
    (C) groove of greater petrosal nerve
    (D) jugular foramen
    (E) mastoid foramen

12. The foramina in the cribriform plate transmit which of the following?
    (A) axons of olfactory cells
    (B) posterior ethmoidal arteries
    (C) ophthalmic arteries
    (D) dural veins
    (E) sympathetic plexus

13. In addition to the optic nerves, the optic canals transmit which of the following?
    (A) ophthalmic veins
    (B) oculomotor nerve
    (C) internal carotid artery
    (D) ophthalmic arteries
    (E) nerve branches to the meninges

14. The maxillary division of the trigeminal nerve is transmitted through which of the following?
    (A) foramen rotundum
    (B) foramen ovale
    (C) superior orbital fissure
    (D) foramen spinosum
    (E) foramen lacerum

15. Which of the following transmits a nerve branch to the meninges?
    (A) foramen spinosum
    (B) mastoid foramen
    (C) jugular foramen
    (D) foramen magnum
    (E) foramen oval

16. The foramen magnum transmits all of the following EXCEPT
    (A) medulla and meninges
    (B) vertebral arteries
    (C) spinal roots of the accessory nerve
    (D) dural veins
    (E) internal carotid artery

17. The jugular foramen transmits all of the following EXCEPT
    (A) glossopharyngeal nerve
    (B) vagus nerve
    (C) accessory nerve
    (D) inferior petrosal and sigmoid sinuses
    (E) sympathetic plexus

18. Which of the following is NOT true regarding the buccinator?
    (A) It is a muscle of mastication.
    (B) It is innervated by the facial nerve.
    (C) It presses the cheek against the molar teeth to assist in chewing.
    (D) It expels air from the oral cavity.
    (E) It draws the mouth to one side when acting unilaterally.

19. Which of the following is NOT a muscle of facial expression?

(A) platysma
(B) nasalis
(C) frontal belly of occiptofrontalis
(D) orbicularis oculi
(E) temporalis

20. The mentalis does which of the following?

(A) elevates eyebrows and skin of forehead
(B) functions as a sphincter of oral opening
(C) elevates lip upper limb and dilates nostril
(D) elevates and protrudes lower lip
(E) draws ala of nose toward nasal septum

21. The orbicularis oculi is innervated by

(A) optic nerve
(B) oculomotor nerve
(C) trochlear nerve
(D) trigeminal nerve
(E) facial nerve

22. The facial nerve innervates all of the following EXCEPT

(A) procerus
(B) corrugator supercilii
(C) masseter
(D) depressor anguli oris
(E) zygomaticus major

23. All muscles of facial expression develop from which pharyngeal arch?

(A) first arch
(B) second arch
(C) third arch
(D) fourth arch
(E) sixth arch

24. Which of the following does NOT insert on the angle of the mouth?

(A) platysma
(B) buccinator
(C) risorius
(D) zygomaticus major
(E) depressor septi

25. In respect to the orbicularis oculi, which of the following is true?

(A) Its orbital part delicately closes the eyelids in blinking.
(B) Its palpebral part draws the eyelids medially so that tears may be drained.
(C) Its lacrimal part tightly closes the eyelids in squinting.
(D) It is innervated by a zygomatic branch of the facial nerve.
(E) It takes origin from the skin of the margin of the orbit and the tarsal plate, and it inserts on the lacrimal bone, medial palpebral ligament, and medial orbital margin.

26. Which of the following cutaneous nerves is derived from the ophthalmic nerve?

(A) external nasal nerve
(B) infraorbital nerve
(C) zygomaticotemporal nerve
(D) zygomaticofacial nerve
(E) auriculotemporal nerve

27. Which of the following nerves arises by two roots that surround the middle meningeal artery?

(A) auriculotemporal nerve
(B) buccal nerve
(C) mental nerve
(D) zygomaticotemporal nerve
(E) zygomaticofacial nerve

28. Which of the following nerves is NOT correctly matched with its distribution?

(A) frontal nerve . . . skin of forehead, scalp, eyelid, and nose
(B) supraorbital nerve . . . skin of forehead as far as vertex
(C) supratrochlear . . . skin in middle of forehead
(D) infratrochlear . . . skin and conjunctiva of upper eyelid
(E) lacrimal . . . skin on dorsum of nose

29. Which of the following sequences of nerve branches is NOT correct?

(A) ophthalmic nerve . . . frontal nerve . . . supraorbital nerve
(B) ophthalmic nerve . . . frontal nerve . . . infratrochlear nerve
(C) maxillary nerve . . . infraorbital nerve
(D) mandibular nerve . . . auriculotemporal nerve
(E) mandibular nerve . . . inferior alveolar nerve . . . mental nerve

30. Which of the following nerves is correctly described with respect to its course?

(A) The auriculotemporal travels from the anterior division of the mandibular nerve in the infratemporal fossa to reach the cheek.
(B) The buccal nerve travels from the posterior division of the mandibular nerve between the neck of the mandible and the external acoustic meatus.
(C) The lacrimal nerve passes through the palpebral fascia of the upper eyelid near the lateral canthus of the eye.
(D) The infratrochlear nerve passes superiorly on the medial surface of the supraorbital nerve.
(E) The supratrochlear nerve emerges through the supraorbital notch and divides into branches.

31. The lacrimal nerve innervates the lacrimal gland with fibers

(A) from the ophthalmic nerve
(B) borrowed via a communicating branch from the maxillary nerve
(C) borrowed from the mandibular nerve
(D) from the optic nerve
(E) from the sympathetic plexus

32. The maxillary nerve gives off branches to which of the following ganglia?

(A) otic ganglion
(B) ciliary ganglion
(C) submandibular ganglion
(D) pterygopalatine ganglion
(E) geniculate ganglion

33. Which of the following is NOT a branch of the facial nerve?

(A) temporal
(B) zygomatic
(C) buccal
(D) mental
(E) cervical

34. The temporal branch of CN VII does NOT innervate which of the following?

(A) auricularis superior
(B) auricularis anterior
(C) occipitofrontalis (frontal belly)
(D) orbicularis oculi (superior part)
(E) orbicularis oculi (inferior part)

35. Which of the following is true regarding the stylomastoid foramen?

(A) It is located between the styloid and mastoid processes of the sphenoid bone.
(B) CN V is transmitted through it.
(C) Sensory nerves of the face travel through it.
(D) The stylomastoid branch of the posterior auricular artery travels though it.
(E) It is a common site of lesion for the glossopharyngeal nerve.

36. Which of the following is NOT a branch of the facial artery?

(A) inferior labial artery
(B) superior labial artery
(C) lateral nasal artery
(D) angular artery
(E) retromandibular artery

37. Which of the following is a branch of internal carotid artery?

(A) supratrochlear artery
(B) facial artery
(C) posterior auricular artery

(D) superficial temporal artery
(E) mental artery

38. Which of the following is NOT true?

(A) The angular artery is distributed to the superior part of the cheek and lower eyelid.
(B) The occipital artery is distributed to the scalp in the back of the head.
(C) The superficial temporal artery is distributed to the parotid gland and duct.
(D) The mental artery is distributed to facial muscles and skin of the chin.
(E) The supratrochlear artery is distributed to the muscles and skin of the scalp.

39. Which of the following is NOT contained within the parotid gland?

(A) facial nerve
(B) facial artery
(C) retromandibular vein
(D) parotid lymph nodes
(E) branches of the facial nerve

40. Parasympathetic fibers from CN IX travel to the parotid gland via which of the following nerves?

(A) auriculotemporal nerve
(B) great auricular nerve
(C) directly from the glossopharyngeal nerve
(D) external carotid nerve plexus
(E) retromandibular nerve

41. Parasympathetic fibers from CN IX synapse in which ganglion before traveling to the parotid gland?

(A) otic ganglion
(B) ciliary ganglion
(C) submandibular ganglion
(D) pterygopalatine ganglion
(E) trigeminal ganglion

42. Which of the following lists the layers of the scalp in the correct order?

(A) skin, connective tissue, auricular layer, loose connective tissue, pericranium
(B) skin, aponeurosis, connective tissue, loose connective tissue, pericranium
(C) skin, connective tissue, aponeurosis, loose connective tissue, paradural layer
(D) skin, connective tissue, aponeurosis, loose connective tissue, pericranium
(E) skin, cartilaginous layer, aponeurosis, loose connective tissue, pericranium

43. Which of the following descriptions is correct?

(A) The superior sagittal sinus runs superior to the brain in the inferior free border of the cerebral falx and ends in the straight sinus.
(B) The inferior sagittal sinus is formed by the union of the superior sagittal sinus and the great cerebral vein and ultimately joins the confluence of the sinuses.
(C) The transverse sinuses follow S-shaped courses in the posterior cranial fossa and ultimately become the internal jugular veins.
(D) The occipital sinus lies in the convex attached border of the cerebral falx, running from the crista galli to the internal occipital protuberance.
(E) The cavernous sinus is situated on each side of the sella turcica.

44. The cavernous sinus receives blood from all of the following EXCEPT

(A) superior and inferior petrosal sinuses
(B) superior ophthalmic veins
(C) inferior ophthalmic veins
(D) superficial middle cerebral vein
(E) sphenoparietal sinus

45. Which of the following nerves is NOT contained within the cavernous sinus?

(A) optic nerve
(B) oculomotor nerve
(C) trochlear nerve
(D) trigeminal nerve (specifically the ophthalmic division)
(E) abducent nerve

46. Which of the following does NOT contribute to innervation of the dura mater?

(A) ophthalmic division of the trigeminal nerve
(B) maxillary division of the trigeminal nerve
(C) mandibular division of the trigeminal nerve
(D) facial nerve
(E) C1, C2, and C3

47. Which of the following is correct in respect to the brain?

(A) The midbrain is composed of the epithalamus, dorsal thalamus, and hypothalamus and surrounds the third ventricle.
(B) The pons is the rostral part of the brainstem and lies at the junction of the middle and posterior cranial fossae.
(C) The diencephalon lies dorsal to the pons and medulla and ventral to the posterior part of the cerebrum, beneath the tentorium.
(D) The cavity of the medulla oblongata forms the inferior part of the fourth ventricle.
(E) The cerebrum occupies the middle and posterior cranial fossae and houses the third and fourth ventricles.

48. Which of the following correctly describes the flow of cerebrospinal fluid?

(A) lateral ventricles . . . cerebral aqueduct . . . 3rd ventricle . . . interventricular foramen . . . 4th ventricle . . . median and lateral apertures . . . subarachnoid space
(B) lateral ventricles . . . interventricular foramen . . . 3rd ventricle . . . cerebral aqueduct . . . 4th ventricle . . . median and lateral apertures . . . subarachnoid space
(C) lateral ventricles . . . interventricular foramen . . . 3rd ventricle . . . median and lateral apertures . . . 4th ventricle . . . cerebral aqueduct . . . subarachnoid space
(D) lateral ventricles . . . median and lateral apertures . . . 3rd ventricle . . . cerebral aqueduct . . . 4th ventricle . . . interventricular foramen . . . subarachnoid space
(E) lateral ventricles . . . straight sinus . . . 3rd ventricle . . . confluence of the sinuses . . . 4th ventricle . . . median and lateral apertures . . . subarachnoid space

49. Which of the following is true regarding the carotid canal?

(A) The carotid canal is located in the inferior surface of the sphenoid bone in the middle cranial fossa.
(B) Structures actually pass across rather than through the area of the carotid canal, which is an artifact in dry skulls and is actually closed by cartilage in life.
(C) The carotid canal contains the internal carotid artery with associated parasympathetic nerves.
(D) The carotid canal contains the internal carotid venous plexus connecting the cavernous sinus and the internal jugular vein.
(E) The greater petrosal nerve enters behind and above the carotid canal and leaves anteriorly as the nerve of the pterygoid canal.

50. Which of the following foramina is unpaired?

(A) foramen lacerum
(B) greater palatine foramen
(C) foramen cecum
(D) lesser palatine foramen
(E) pterygoid canal

51. The oculomotor nerve emerges between which two arteries of the cerebral arterial circle?

(A) posterior inferior cerebellar artery and anterior inferior cerebellar artery
(B) anterior inferior cerebellar artery and labyrinthine artery
(C) labyrinthine artery and superior cerebellar artery
(D) superior cerebellar artery and posterior cerebral artery
(E) posterior cerebral artery and middle cerebral artery

52. What nerve emerges between the labyrinthine artery and the anterior inferior cerebellar artery?

(A) optic
(B) trochlear
(C) trigeminal
(D) abducent
(E) facial

53. Which artery is NOT a branch of the vertebral artery system?

(A) basilar
(B) posterior cerebral
(C) posterior communicating
(D) anterior spinal
(E) ophthalmic

54. Which of the following arteries is correctly matched with its distribution?

(A) anterior cerebral . . . temporal and occipital lobes of brain
(B) middle cerebral . . . inferior aspect of cerebral hemispheres and occipital lobe
(C) middle meningeal . . . calvaria
(D) posterior cerebral . . . brainstem and cerebellum
(E) basilar . . . optic tract, cerebral peduncle, internal capsule, and thalamus

55. Which of the following lists best describes the pathway of tears from the lacrimal glands to the nasal cavity?

(A) lacrimal ducts . . . lacrimal lake . . . lacrimal canaliculi . . . lacrimal sac . . . nasolacrimal duct
(B) lacrimal ducts . . . lacrimal sac . . . lacrimal canaliculi . . . lacrimal lake . . . nasolacrimal duct
(C) lacrimal canaliculi . . . lacrimal lake . . . lacrimal ducts . . . lacrimal sac . . . nasolacrimal duct
(D) lacrimal canaliculi . . . lacrimal sac . . . lacrimal ducts . . . lacrimal lake . . . nasolacrimal duct
(E) lacrimal punctum . . . lacrimal lake . . . lacrimal papilla . . . lacrimal sac . . . nasolacrimal duct

56. Which of the following best describes the pathway of parasympathetic fibers to the lacrimal gland?

(A) CN VII . . . greater petrosal nerve . . . nerve of the pterygoid canal . . . pterygopalatine ganglion . . . zygomatic branch of V_2 . . . lacrimal branch of V_1
(B) CN IX . . . lesser petrosal nerve . . . nerve of the pterygoid canal . . . pterygopalatine ganglion . . . infraorbital branch of V_2 . . . lacrimal branch of V_1
(C) CN VII . . . greater petrosal nerve . . . otic ganglion . . . infraorbital branch of V_2 . . . lacrimal branch of V_1
(D) CN VII . . . deep petrosal nerve . . . nerve of the pterygoid canal . . . pterygopalatine ganglion . . . infratrochlear branch of V_2 . . . lacrimal branch of V_1
(E) CN V . . . ophthalmic branch of V_1 . . . lacrimal branch of V_1

57. Which of the following muscles does NOT take its origin from the common tendinous ring?

(A) superior rectus
(B) inferior rectus
(C) lateral rectus
(D) medial rectus
(E) superior oblique

58. Which of the following muscles is NOT innervated by the oculomotor nerve?

(A) levator palpebrae superioris
(B) lateral rectus
(C) medial rectus
(D) inferior oblique
(E) superior rectus

59. Which of the following muscles is NOT properly matched with it main action?

(A) lateral rectus . . . abducts eyeball
(B) superior rectus . . . elevates, adducts, and rotates eyeball medially
(C) inferior rectus . . . depresses, adducts, and rotates eyeball medially
(D) superior oblique . . . adducts, elevates, and rotates eyeball laterally
(E) inferior oblique . . . abducts, elevates, and rotates eyeball laterally

60. Which of the following nerves is correctly matched with its distribution?

(A) long ciliary . . . postsynaptic sympathetic fibers to the dilator pupillae
(B) short ciliary . . . parasympathetic and sympathetic fibers to lens and cornea
(C) frontal . . . conjunctiva and lacrimal gland
(D) infratrochlear . . . mucous membrane of sphenoidal and ethmoid sinuses
(E) ethmoidal . . . conjunctiva and eyelids

61. Which of the following is true in respect to the ciliary ganglion?

(A) Sympathetic fibers synapse in the ciliary ganglion.
(B) Afferent fibers from the iris and cornea pass through the ganglion.
(C) The ganglion is located between the optic nerve and medial rectus.
(D) Parasympathetic fibers in the ganglion are derived from CN VII.
(E) Parasympathetic fibers in the ganglion are distributed to the retina and lens.

62. Which of the following arteries is NOT a branch of the ophthalmic artery?

(A) supraorbital
(B) supratrochlear
(C) lacrimal
(D) anterior ethmoidal
(E) infraorbital

63. Which of the following arteries is correctly paired with its course and distribution?

(A) central artery of retina . . . runs adjacent to optic nerve, supplying rods and cones
(B) lacrimal artery . . . runs along medial rectus to supply lacrimal gland and frontal sinus
(C) short posterior cilaries . . . pierce sclera to supply choroid, rods, and cones
(D) long posterior cilaries . . . supplies lens and cornea
(E) posterior ethmoidal . . . supplies dorsal aspect of nose

64. Which of the following is NOT contained in the infratemporal fossa?

(A) parts of temporal, lateral pterygoid, and medial pterygoid muscles
(B) maxillary artery
(C) pterygoid venous plexus
(D) mandibular, inferior alveolar, buccal, and lingual nerves
(E) pterygopalatine ganglion

65. Which of the following muscles is NOT a muscle of mastication?

(A) buccinator
(B) temporalis
(C) medial pterygoid
(D) lateral pterygoid
(E) masseter

66. The muscles of mastication are associated with which branchial arch?

(A) first arch
(B) second arch
(C) third arch
(D) fourth arch
(E) fifth arch

67. Which of the following depresses the mandible?

(A) lateral pterygoid
(B) medial pterygoid
(C) temporalis
(D) masseter
(E) mylohyoid

68. Which of the following is NOT a branch of the first (mandibular) part of the first part of the maxillary artery?

(A) deep auricular
(B) anterior tympanic
(C) middle meningeal and accessory meningeal
(D) inferior alveolar
(E) descending palatine

69. Which of the following is NOT a branch of the second (pterygoid) part of the maxillary artery?

(A) deep temporal
(B) labyrinthine
(C) pterygoid
(D) masseteric
(E) buccal

70. Which of the following branches of the third (pterygopalatine) part of the maxillary artery is correctly paired with its distribution?

(A) infraorbital . . . maxillary molar and premolar teeth, lining of maxillary sinus, gingival
(B) posterior superior alveolar . . . inferior eyelid, lacrimal sac, side of nose, superior lip
(C) pharyngeal . . . maxillary gingiva, palatine glands, mucous membrane of roof of mouth
(D) artery of pterygoid canal . . . superior part of pharynx, auditory tube, tympanic cavity
(E) descending palatine . . . roof of pharynx, sphenoidal sinus, inferior part of auditory tube

71. Which of the following is NOT true in respect to the sphenopalatine artery?

(A) It is the termination of the maxillary artery.
(B) It supplies the lateral nasal wall.
(C) It supplies the nasal septum.
(D) It supplies the paranasal sinuses.
(E) It is transmitted through the incisive foramen.

72. Which of the following is true in respect to the otic ganglion?

(A) It is located in the infratemporal fossa just inferior to the foramen rotundum.
(B) Presynaptic parasympathetic fibers in the ganglion are derived from the facial nerve.
(C) Postsynaptic parasympathetic fibers in the ganglion are destined for the parotid gland.
(D) Sympathetic fibers synapse in the ganglion before continuing on to sweat glands, erector pili muscles, and blood vessels.
(E) The ganglion contains cell bodies for fibers of the trigeminal nerve.

73. Which nerve is NOT correctly matched with its distribution?

(A) lingual nerve . . . sensation from the anterior two-thirds of the tongue
(B) lingual nerve . . . sensation from floor of mouth and lingual gingivae
(C) chorda tympani nerve . . . taste fibers from anterior two-thirds of tongue
(D) chorda tympani nerve . . . secretomotor fibers to submandibular and sublingual glands
(E) chorda tympani nerve . . . motor fibers to tensor tympani

74. The temporomandibular joint is what type of joint?

(A) fibrous joint
(B) cartilaginous joint
(C) modified hinge-type synovial joint
(D) pivot-type synovial joint
(E) saddle-type synovial joint

75. Which of the following is correct in respect to the hard palate?

(A) The hard palate is composed primarily of the maxillary bones.
(B) The incisive canal and foramen transmit the incisive nerves and greater palatine vessels.
(C) The greater palatine foramen transmits the nasopalatine nerves and greater palatine nerve.
(D) The lesser palatine foramina transmit the lesser palatine nerves but not the lesser palatine vessels.
(E) The descending palatine artery is a branch of the internal carotid artery.

76. Which of the following palate muscles is NOT innervated by the cranial part of the accessory nerve through a pharyngeal branch of the vagus nerve via the pharyngeal plexus?

(A) tensor veli palatini
(B) levator veli palatini
(C) palatoglossus
(D) palatopharyngeus
(E) musculus uvulae

77. Which of the following is correctly matched with its action?

(A) musculus uvulae . . . pulls uvula inferiorly
(B) palatopharyngeus . . . pulls walls of pharynx inferiorly, posteriorly, and laterally during swallowing
(C) palatoglossus . . . depresses posterior part of tongue and draws soft palate away from tongue
(D) levator veli palatini . . . depresses soft palate during swallowing and yawning
(E) tensor veli palatini . . . opens auditory tube during swallowing and yawning

78. Which of the following is NOT a type of lingual papilla?

(A) vallate papillae
(B) foliate papillae
(C) bacilliform papillae
(D) filiform papillae
(E) fungiform papillae

79. Which of the following muscles is NOT innervated by the hypoglossal nerve?

(A) genioglossus
(B) hyoglossus
(C) styloglossus
(D) palatoglossus
(E) intrinsic muscles of the tongue

80. Which of the following tongue muscles is correctly paired with its action?

(A) superior and inferior longitudinal . . . curls tip and sides of tongue and shortens tongue
(B) transverse . . . flattens and broadens tongue
(C) vertical . . . narrows and elongates tongue
(D) genioglossus and hyoglossus . . . elevates posterior part of tongue
(E) palatoglossus . . . depresses and retracts tongue

81. Which of the following is true in respect to innervation of the tongue?

(A) The chorda tympani nerve innervates the mucosa of the anterior two-thirds of the tongue in respect to general sensation (touch and temperature).
(B) The chorda tympani nerve innervates the anterior two-thirds of the tongue in respect to special sensation (taste).
(C) The lingual nerve innervates the posterior one-third of the tongue in respect to general sensation (touch and temperature).
(D) The lingual nerve innervates the posterior one-third of the tongue in respect to special sensation (taste).
(E) The glossopharyngeal nerve innervates the area of the tongue just anterior to the epiglottis in respect to both general and special sensation.

82. How do parasympathetic and taste fibers from the chorda tympani reach their destination?

(A) via the lingual nerve, a branch of the mandibular division of the trigeminal nerve
(B) via the lingual nerve, a branch of the glossopharyngeal nerve
(C) via the lingual nerve, a branch of the vagus nerve
(D) via intermingled fibers in the otic ganglion
(E) via the submandibular nerve, a branch of the hypoglossal nerve

83. Which of the following taste sensations is correctly paired with its tongue region?

(A) savoriness . . . posterior part
(B) sourness . . . apex
(C) bitterness . . . apex
(D) saltiness . . . lateral margins
(E) sweetness . . . posterior part

84. Which of the following vessels is NOT correctly paired with its respective area of supply or drainage?

(A) dorsal lingual arteries . . . supply submandibular gland
(B) deep lingual artery . . . supplies anterior tongue
(C) sublingual artery . . . supplies sublingual gland and floor of mouth
(D) dorsal lingual veins . . . accompany the lingual artery
(E) deep lingual veins . . . drain the apex of the tongue, joining the sublingual vein

85. Which of the following is NOT correct?

(A) The parotid gland is supplied by branches of the external carotid and superficial temporal arteries.
(B) The submandibular gland is supplied by the submental artery.
(C) The submandibular gland is innervated by the parasympathetic fibers of the glossopharyngeal nerve that synapsed in the submandibular ganglion.
(D) The sublingual glands are supplied by the sublingual and submental arteries.
(E) The sublingual glands are innervated by parasympathetic fibers of the facial nerve.

86. Which of the following is NOT an opening to the pterygopalatine fossa?

(A) pterygomaxillary fissure
(B) sphenopalatine foramen
(C) inferior orbital fissure
(D) foramen rotundum
(E) foramen ovale

87. Which of the following is NOT contained in the pterygopalatine fossa?

(A) third part of maxillary artery
(B) maxillary nerve
(C) nerve of the pterygoid canal
(D) pterygopalatine ganglion
(E) optic nerve

88. Which of the following foramina are NOT properly matched with the structures they transmit?

(A) inferior orbital fissure . . . ophthalmic nerve, infraorbital vessels, orbital branches of pterygopalatine ganglion
(B) infraorbital foramen and canal . . . infraorbital nerve and vessels
(C) palatovaginal canal (pharyngeal) canal . . . pharyngeal nerves from maxillary nerve and pterygopalatine ganglion and pharyngeal branch of maxillary artery
(D) zygomaticofacial foramen . . . zygomaticofacial nerve and vessels
(E) zygomaticotemporal foramen . . . zygomaticotemporal nerve and vessels

89. The nerve of the pterygoid canal is composed of which of the following?

(A) lesser petrosal nerve and deep petrosal nerve
(B) greater petrosal nerve and deep petrosal nerve
(C) greater petrosal nerve and lesser petrosal nerve
(D) maxillary nerve and deep petrosal nerve
(E) maxillary nerve and greater petrosal nerve

90. The nerve of the pterygoid canal does NOT innervate which of the following?

(A) lacrimal gland
(B) palatine glands
(C) mucosal glands of nasal cavity
(D) mucosal glands of upper pharynx
(E) submandibular gland

91. Which of the following paranasal sinuses communicates with the nasal cavity in the superior meatus?

(A) posterior ethmoidal sinuses
(B) frontal sinus
(C) middle ethmoidal sinuses
(D) sphenoidal sinus
(E) maxillary sinus

92. Where does the nasolacrimal duct communicate with the nasal cavity?

(A) superior meatus
(B) middle meatus
(C) inferior meatus
(D) nasopharynx
(E) sphenoidal sinus

93. Which of the following is a separate bone?

(A) superior nasal concha
(B) middle nasal concha
(C) inferior nasal concha
(D) crista galli
(E) glabella

94. Which of the following does NOT supply the medial and lateral walls of the nasal cavity?

(A) sphenopalatine artery
(B) anterior and posterior ethmoidal arteries
(C) greater palatine artery
(D) infraorbital artery
(E) superior labial artery

95. Which of the following does NOT innervate the nasal mucosa?

(A) sphenopalatine nerve
(B) nasopalatine nerve
(C) greater palatine nerve
(D) anterior ethmoidal nerve
(E) posterior ethmoidal nerve

96. Which of the following is NOT correct regarding innervation of the ear?

(A) The auricle is innervated by the great auricular nerve and auriculotemporal nerve.
(B) The external surface of the tympanic membrane is innervated by the auriculotemporal nerve and even a small branch of the vagus.
(C) The pharyngotympanic tube is innervated by the tympanic plexus (fibers from the facial and glossopharyngeal nerves).
(D) The internal surface of the tympanic membrane is innervated by the glossopharyngeal nerve.
(E) Sensory cell bodies of the vestibulocochlear nerve are located in the geniculate ganglion.

97. Which of the following correctly describes a wall of the tympanic cavity and its underlying structure?

(A) tegmental roof . . . cochlea, contained in the promontory
(B) floor . . . superior bulb of internal jugular vein
(C) medial wall . . . dura mater of the middle cranial fossa
(D) anterior wall . . . mastoid cells and facial nerve
(E) posterior wall . . . carotid canal

98. Which of the following is NOT contained in the tympanic cavity?

(A) auditory ossicles
(B) tympanic plexus
(C) chorda tympani nerve
(D) lesser petrosal nerve
(E) stapedius and tensor tympani muscles

99. Which of the following is NOT true in respect to the pharyngotympanic tube?

(A) The salpingopharyngeus closes the tube.
(B) It opens posterior to the inferior meatus of the nasal cavity.
(C) It equalizes pressure in the middle ear with atmospheric pressure.
(D) The tensor veli palatini and levator veli palatini work together to open the tube.
(E) It is supplied by the ascending pharyngeal artery, middle meningeal artery, and artery of the pterygoid canal.

100. What innervates the tensor tympani?

(A) maxillary nerve
(B) mandibular nerve
(C) chorda tympani
(D) vestibulocochlear nerve
(E) vagus

101. Which of the following is true?

(A) The malleus articulates with the stapes and is moved by the tensor tympani.
(B) The round window is an opening in the medial wall of the tympanic cavity leading to the vestibule of the inner ear and is closed by the base of the stapes.
(C) The tensor tympani assists in preventing damage to the internal ear when one hears loud noises.
(D) The stapedius is innervated by the chorda tympani.
(E) The stapedius pulls the stapes anteriorly and flattens its base, loosening the annular ligament and increasing oscillatory range.

102. Which of the following foramina is NOT correctly matched with its function?

(A) cochlear aqueduct . . . allows bony labyrinth to communicate with subarachnoid space; also contains labyrinthine vein
(B) aqueduct of the vestibule . . . transmits endolymphatic duct, an artery, and a vein
(C) internal acoustic meatus . . . transmits vestibulocochlear nerve and vestibular artery
(D) mastoid canaliculus . . . transmits auricular branch of the vagus
(E) tympanic canaliculus . . . tympanic branch of the glossopharyngeal nerve

103. Which of the following is true?

(A) The cochlear labyrinth is composed of the utricle, the saccule, and three semicircular canals.
(B) The membranous labyrinth contains perilymph.
(C) The basilar membrane secures the cochlear duct to the spiral canal of the cochlea.
(D) The spiral membrane forms the floor of the cochlear duct.
(E) The saccule is continuous with the cochlear duct through the ductus reuniens.

104. Which of the following is NOT true in respect to typical cervical vertebrae?

(A) They have short, bifid spinous processes.
(B) The inferior facets of articular processes are directed inferoposteriorly, and superior facets are directed superoposteriorly.
(C) The transverse processes contain a foramen transversarium, which transmits the vertebral vein and artery except for C7, where it transmits only the vertebral artery.
(D) The vertebral foramen is large and triangular.
(E) The vertebral body is small, with a concave superior surface and a convex inferior surface.

105. Which fascial layer is NOT correctly matched with the structures it encloses?

(A) superficial cervical fascia . . . platysma
(B) investing layer of deep cervical fascia . . . trapezius and sternocleidomastoid
(C) pretracheal layer of deep cervical fascia . . . suprahyoid muscles
(D) prevertebral layer of deep cervical fascia . . . longus colli, longus capitis, scalenes, deep cervical muscles
(E) carotid sheath . . . carotid arteries, internal jugular vein, vagus

106. Which of the following is NOT a superior attachment of the trapezius?

(A) lateral surface of mastoid process
(B) medial third of superior nuchal line
(C) external occipital protuberance
(D) ligamentum nuchae
(E) spinous processes of C7–T12

107. Which of the following is NOT correct in respect to the posterior triangle of the neck?

(A) Its anterior boundary is formed by the posterior border of the SCM.
(B) Its posterior boundary is formed by the anterior border of the trapezius.
(C) Its inferior boundary is formed by the middle third of the clavicle.
(D) Its roof is formed by the platysma.
(E) Its floor is formed by the muscles covered by the prevertebral layer of deep cervical fascia.

108. The anterior triangle of the neck does NOT contain which of the following smaller triangles?

(A) supraclavicular triangle
(B) submandibular triangle
(C) submental triangle
(D) carotid triangle
(E) muscular triangle

109. Which of the following muscles is NOT contained in the posterior cervical triangle?

(A) splenius capitis
(B) levator scapulae
(C) middle scalene
(D) posterior scalene
(E) stylohyoid

110. Which of the following is a suprahyoid muscle?

(A) mylohyoid
(B) sternohyoid
(C) omohyoid
(D) sternothyroid
(E) thyrohyoid

111. Which of the following is innervated by the trigeminal nerve?

(A) mylohyoid
(B) geniohyoid
(C) stylohyoid
(D) posterior belly of the digastric
(E) thyrohyoid

112. Which of the following is NOT an anterior vertebral muscle?

(A) longus colli
(B) longus capitis
(C) splenius capitis
(D) rectus capitis anterior
(E) rectus capitis lateralis

113. Of the following, which is innervated by the dorsal rami?

(A) splenius capitis
(B) levator scapulae
(C) posterior scalene
(D) middle scalene
(E) anterior scalene

114. Which of the following NEVER exists?

(A) middle thyroid artery
(B) middle thyroid vein
(C) thyroid ima artery
(D) parathyroid veins
(E) anterior jugular vein

115. Which of the following contains PAIRED laryngeal cartilages?

(A) thyroid, cricoid, epiglottic
(B) arytenoid, corniculate, cuneiform
(C) arytenoid, cricoid, epiglottic
(D) corniculate, cricoid, cuneiform
(E) cuneiform, corniculate, epiglottic

116. Which of the following intrinsic laryngeal muscles is NOT innervated by the recurrent laryngeal nerve?

(A) transverse arytenoids
(B) oblique arytenoids

(C) cricothyroid
(D) posterior cricoarytenoid
(E) lateral cricoarytenoid

117. Which of the following is the primary tensor of the vocal fold?

(A) cricothyroid
(B) thyroarytenoid
(C) vocalis
(D) aryepiglottic
(E) posterior cricoarytenoid

118. Which of the following abducts the vocal fold?

(A) vocalis
(B) transverse arytenoids
(C) oblique arytenoids
(D) thyroarytenoid
(E) posterior cricoarytenoid

119. Which of the following is the sensory nerve of the larynx?

(A) external laryngeal nerve
(B) recurrent laryngeal nerve
(C) internal laryngeal nerve
(D) paratracheal nerve
(E) inferior thyroid nerve

120. Which of the following are commonly referred to as the adenoids?

(A) pharyngeal tonsils
(B) submandibular glands
(C) palatine tonsils
(D) lingual tonsils
(E) sublingual glands

121. Which of the following is NOT innervated by the cranial root of the accessory nerve?

(A) middle constrictor
(B) inferior constrictor
(C) palatopharyngeus
(D) salpingopharyngeus
(E) stylopharyngeus

122. Which of the following has an insertion on the pharyngeal tubercle of the occipital bone?

(A) superior constrictor
(B) middle constrictor
(C) palatopharyngeus
(D) salpingopharyngeus
(E) stylopharyngeus

123. Which of the following passes through the gap between the superior constrictor and the skull?

(A) tensor veli palatini
(B) ascending palatine artery
(C) glossopharyngeal nerve
(D) stylohyoid ligament
(E) stylopharyngeus

124. Which of the following passes through the gap between the superior and middle constrictors?

(A) glossopharyngeal nerve
(B) levator veli palatini
(C) internal laryngeal nerve
(D) superior laryngeal artery
(E) superior laryngeal vein

125. Which of the following passes through the gap between the middle and inferior constrictors?

(A) superior laryngeal artery
(B) stylopharyngeus
(C) vagus nerve
(D) recurrent laryngeal nerve
(E) inferior laryngeal artery

126. Which of the following passes through the gap inferior to the inferior constrictor?

(A) vagus nerve
(B) internal laryngeal nerve
(C) superior laryngeal nerve
(D) inferior laryngeal artery
(E) superior laryngeal vein

127. Which of the following is correct?

(A) Le Fort I fracture: horizontal fracture of the maxillae
(B) Le Fort I fracture: fracture through the maxillary sinuses, infraorbital foramina, lacrimals, and ethmoids
(C) Le Fort II fracture: horizontal fracture through superior orbital fissures, ethmoid and nasal bones, and greater wings of the sphenoids
(D) Le Fort III fracture: horizontal fracture of the maxillae
(E) Le Fort III fracture: fracture through the maxillary sinuses, infraorbital foramina, lacrimals, and ethmoids

128. Which of the following is NOT present at birth?

(A) styloid process
(B) mastoid process
(C) external occipital protuberance
(D) tympanic membrane
(E) clavicles

129. The inferior alveolar nerve is best blocked at which location for dental work?

(A) mental foramen
(B) greater palatine foramen
(C) less palatine foramen
(D) mandibular foramen
(E) lingual foramen

130. A lesion to the zygomatic branch of CN VII would cause which of the following?

(A) the inability to empty food from the vestibule of the cheeks
(B) a drooping corner of the mouth
(C) a ringing in the ear
(D) paralysis of the muscles of mastication
(E) a drooping lower eyelid

131. The facial veins make clinically important connections with the cavernous sinus through which veins?

(A) lingual veins
(B) trigeminal veins
(C) superior ophthalmic veins
(D) great cerebral vein
(E) meningeal veins

132. An epidural hematoma consists of blood from which vessel?

(A) middle meningeal artery
(B) cerebral veins
(C) internal carotid artery
(D) circle of Willis
(E) vertebral artery

133. Cerebral compression is NOT attributed to which of the following?

(A) intracranial collections of blood
(B) obstruction of CSF flow
(C) intracranial tumors or abscesses
(D) edema of brain
(E) viral accumulation at blood-brain barrier

134. Ptosis results from a lesion of which nerve?

(A) optic nerve
(B) oculomotor nerve
(C) trochlear nerve
(D) trigeminal nerve
(E) abducens nerve

135. Horner syndrome is caused by a lesion of which of the following?

(A) oculomotor nerve
(B) trigeminal nerve
(C) facial nerve
(D) vagus nerve
(E) cervical sympathetic trunk

136. Which of the following is NOT a symptom of Horner syndrome?

(A) pupillary constriction
(B) ptosis
(C) sinking in of one eye
(D) absence of sweating on face and neck
(E) lack of lacrimation

137. A lesion of the hypoglossal nerve would result in which of the following?

(A) loss of taste on posterior one-third of tongue
(B) deviation of protruded tongue toward unaffected side
(C) deviation of protruded tongue toward affected side
(D) inability to retract tongue
(E) loss of salivation

138. What type of injury or condition might cause a lesion to the olfactory tract?

(A) fracture involving optic canal
(B) fracture of cribriform plate
(C) intracerebral clot in occipital lobe of brain
(D) pituitary tumor
(E) epidural hematoma

139. Sagging of the soft palate, deviation of the uvula to the normal side, and hoarseness might be caused by a lesion to which nerve?

(A) facial nerve
(B) glossopharyngeal nerve
(C) vagus nerve
(D) accessory nerve
(E) hypoglossal nerve

140. A superficial neck laceration might result in which abnormal finding?

(A) paralysis of the scm and superior fibers of the trapezius, drooping of the shoulder
(B) paralysis of the mylohyoid, anterior belly of the digastric, tensor tympani, and tensor veli palatini
(C) paralysis of the posterior belly of the digastric, stylohyoid, and stapedius
(D) anosmia
(E) tinnitus

141. The eye is turned down and out. What type and site of lesion do you expect?

(A) fracture of cribriform plate
(B) stretching of a nerve as it courses around the brainstem
(C) laceration or contusion in the parotid region
(D) pressure from herniating uncus on the nerve or fracture in the cavernous sinus
(E) acoustic neuroma

142. A laceration or contusion in the parotid region or a fracture of the temporal bone might damage which nerve?

(A) trigeminal nerve
(B) abducens nerve
(C) facial nerve
(D) glossopharyngeal nerve
(E) vagus nerve

143. Which nerve might be damaged by a fracture involving the cavernous sinus?

(A) olfactory tract
(B) optic nerve
(C) abducens nerve
(D) facial nerve
(E) vestibulocochlear nerve

144. The stylopharyngeus is associated with which branchial arch?

(A) first
(B) second
(C) third
(D) fourth
(E) sixth

145. The arytenoid and cricoid cartilages and laryngeal connective tissue are formed by what?

(A) lateral plate mesoderm
(B) paraxial mesoderm
(C) neural crest
(D) ectodermal placodes
(E) endoderm

146. Which of the following prominences is NOT correctly paired with the structures formed from it?

(A) frontonasal: forehead, bridge of nose
(B) maxillary: lateral portion of upper lip
(C) medial nasal: philtrum of upper lip, crest and tip of nose
(D) lateral nasal: alae of nose
(E) mandibular: cheeks

DIRECTIONS (Questions 147 through 151): Identify the anatomical features indicated on the art below.

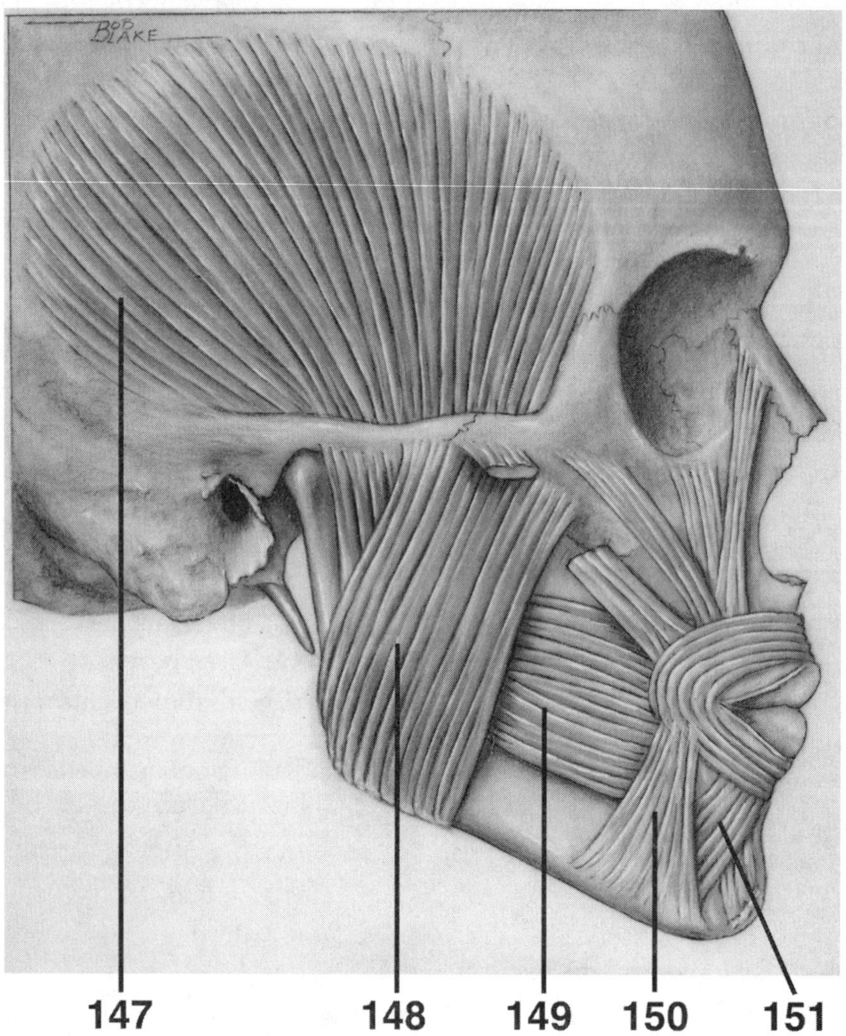

DIRECTIONS (Questions 152 through 156): Identify the anatomical features indicated on the art below.

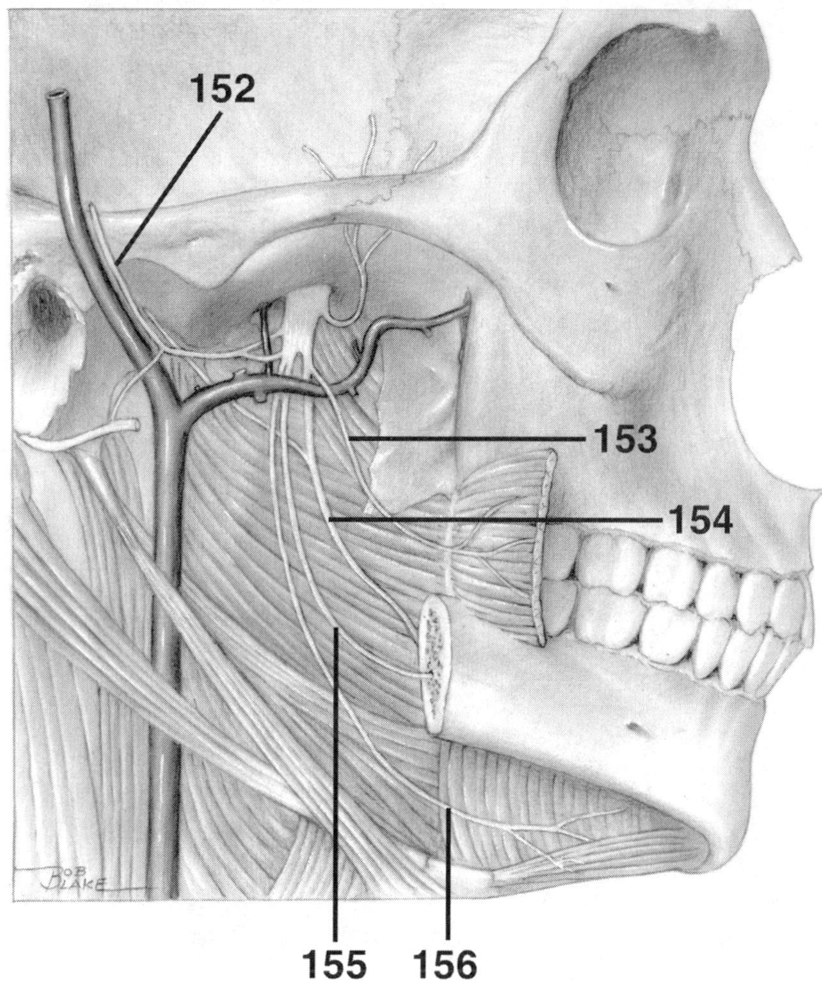

DIRECTIONS (Questions 157 through 161): Identify the anatomical features indicated on the art below.

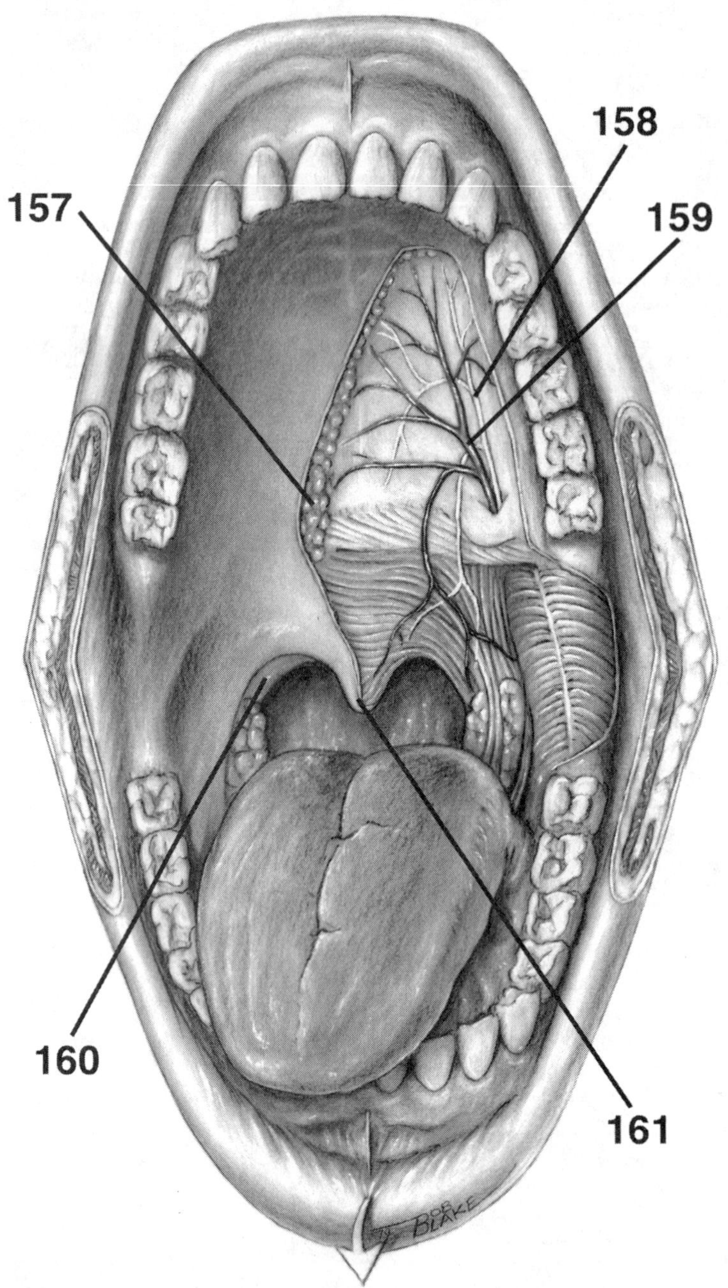

DIRECTIONS (Questions 162 through 166): Identify the anatomical features indicated on the art below.

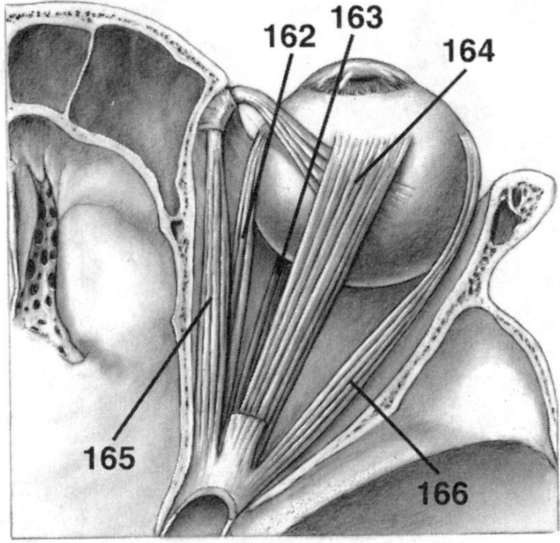

DIRECTIONS (Questions 167 through 171): Identify the anatomical features indicated on the art below.

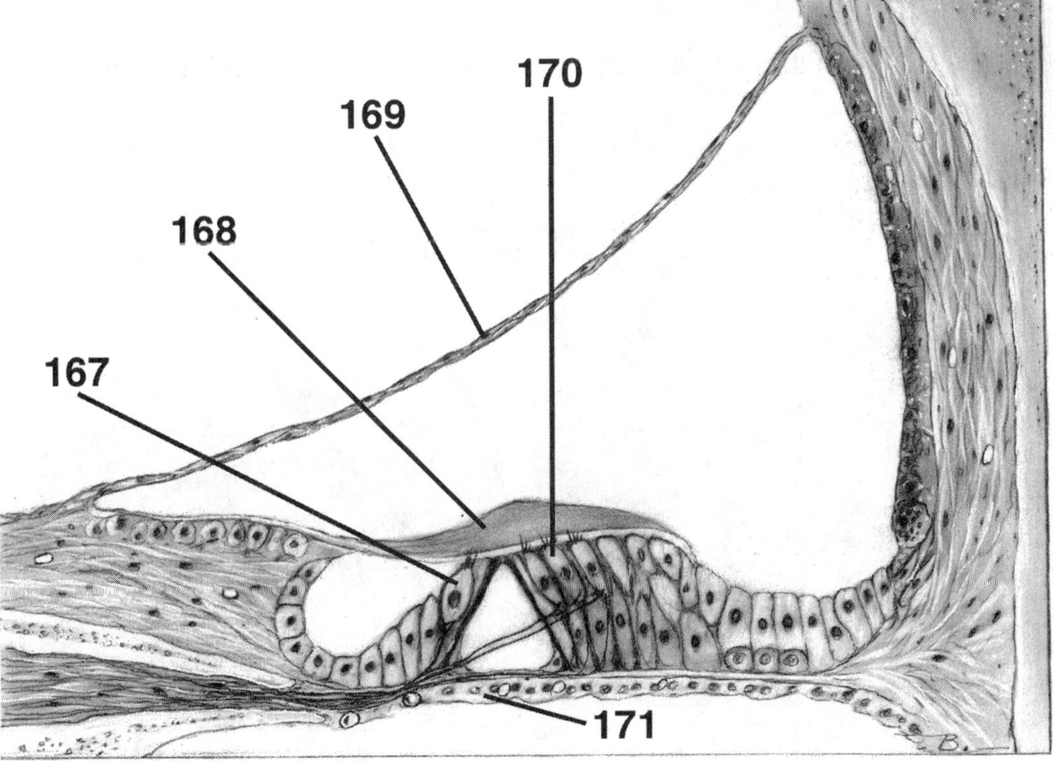

DIRECTIONS (Questions 172 through 176): Identify the anatomical features indicated on the art below.

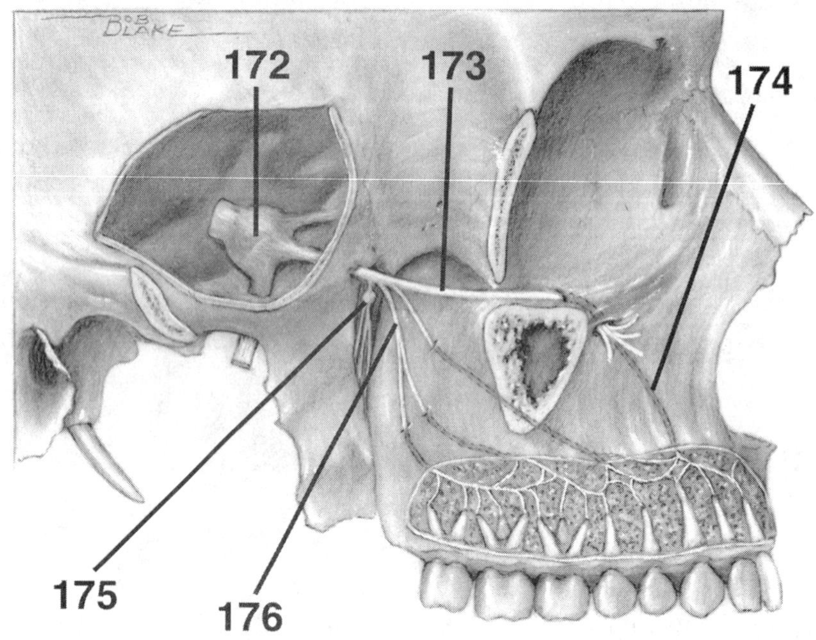

DIRECTIONS (Questions 177 through 181): Identify the anatomical features indicated on the art below.

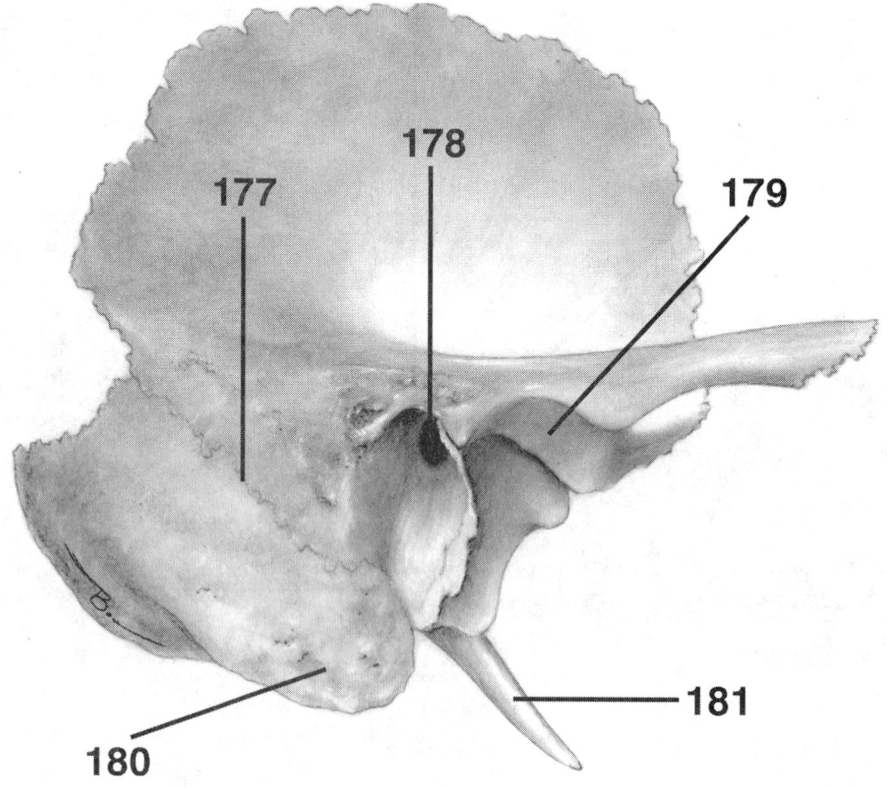

DIRECTIONS (Questions 182 through 186): Identify the anatomical features indicated on the art below.

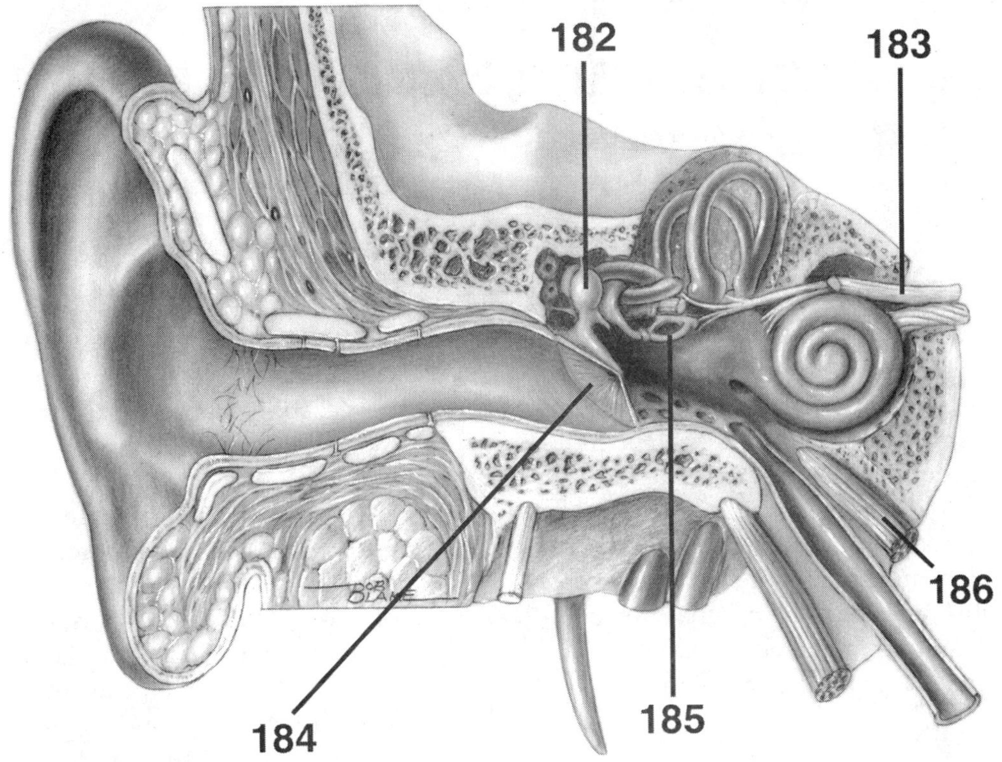

Answers and Explanations

1. **(A)** The bones of the neurocranium include the frontal bone, paired parietal bones, paired temporal bones, the occipital bone, the sphenoid bone, and the ethmoid bone *(Moore, p 832).*

2. **(C)** The bones of the facial skeleton (viscerocranium or splanchnocranium) include the vomer, the mandible, inferior nasal conchae, the palatine bones, the zygomatic bones, the maxillae, the nasal bones, and the lacrimal bones *(Moore, p 832).*

3. **(A)** When the frontal suture persists, it is known as the metopic suture *(Moore, p 834).*

4. **(B)** The external occipital protuberance is also known as the inion *(Moore, p 839).*

5. **(A)** Lambda is the point on the calvaria at the junction of the sagittal and lambdoid sutures *(Moore, p 842).*

6. **(C)** The vertex is the superior point of the neurocranium in the midline *(Moore, p 842).*

7. **(E)** The foramen magnum is in the posterior cranial fossa *(Moore, p 846).*

8. **(A)** The foramen cecum is located in the anterior cranial fossa *(Moore, p 846).*

9. **(E)** The anterior and posterior ethmoidal foramina transmit anterior and posterior ethmoidal arteries and nerves, not emissary veins *(Moore, p 846).*

10. **(B)** The superior orbital fissure transmits the ophthalmic veins, ophthalmic division of the trigeminal nerve, oculomotor nerve, trochlear nerve, abducens nerve, and sympathetic fibers *(Moore, p 846).*

11. **(B)** The foramen ovale transmits the accessory meningeal artery. The foramen spinosum transmits the middle meningeal artery. The groove of the greater petrosal nerve transmits the petrosal branch of the middle meningeal artery. The jugular foramen transmits the meningeal branches of the ascending pharyngeal and occipital arteries. The mastoid foramen transmits the meningeal branch of the occipital artery *(Moore, p 846).*

12. **(A)** The foramina in the cribriform plate transmit axons of olfactory cells in the olfactory epithelium *(Moore, p 846).*

13. **(D)** The optic canals transmit the optic nerves and the ophthalmic arteries *(Moore, p 846).*

14. **(A)** The foramen rotundum transmits the maxillary division of the trigeminal nerve *(Moore, p 846).*

15. **(A)** The foramen spinosum transmits the meningeal branch of the mandibular division of the trigeminal nerve *(Moore, p 846).*

16. **(E)** The foramen magnum transmits the medulla and meninges, vertebral arteries, spinal roots of the accessory nerve, dural veins, and the anterior and posterior spinal arteries *(Moore, p 846).*

17. **(E)** The jugular foramen transmits the glossopharyngeal nerve, vagus nerve, accessory

nerve, internal jugular vein, inferior petrosal and sigmoid sinuses, and meningeal branches of ascending pharyngeal and occipital arteries *(Moore, p 846).*

18. **(A)** The buccinator, a muscle of facial expression, is innervated by the facial nerve. It presses the cheek against the molar teeth to assist in chewing, and it expels air from the oral cavity. It also draws the mouth to one side when acting unilaterally *(Moore, p 851).*

19. **(E)** The temporalis is a muscle of mastication, not of facial expression *(Moore, p 851).*

20. **(D)** The mentalis elevates and protrudes the lower lip *(Moore, p 851).*

21. **(E)** The facial nerve innervates the orbicularis oculi *(Moore, p 851).*

22. **(C)** The trigeminal nerve innervates the masseter *(Moore, p 921).*

23. **(B)** The muscles of facial expression develop from the second pharyngeal arch *(Moore, p 852).*

24. **(E)** The depressor septi inserts into part of the nasal septum and widens the alar part of the nasalis muscle during deep inspiration *(Moore, p 856).*

25. **(D)** The orbicularis oculi is innervated by a zygomatic branch of the facial nerve *(Moore, pp 851, 855–856).*

26. **(A)** The external nasal nerve is a cutaneous nerve derived from the ophthalmic nerve *(Moore, p 857).*

27. **(A)** The auriculotemporal nerve arises by two roots that surround the middle meningeal artery *(Moore, p 861).*

28. **(E)** The lacrimal nerve is distributed to the lacrimal gland and a small area of skin as well as part of the conjunctiva. The external nasal nerve is distributed to skin on the dorsum of the nose, including the tip of the nose *(Moore, p 860).*

29. **(B)** The ophthalmic nerve gives rise to the nasociliary nerve, which in turn gives rise to the infratrochlear nerve *(Moore, p 860).*

30. **(C)** The lacrimal nerve passes through the palpebral fascia of the upper eyelid near the lateral canthus of the eye *(Moore, p 860).*

31. **(B)** The lacrimal nerve innervates the lacrimal gland with fibers borrowed via a communicating branch from the maxillary nerve *(Moore, p 859).*

32. **(D)** The maxillary nerve gives off branches to the pterygopalatine ganglion *(Moore, p 859).*

33. **(D)** The mental nerve is a branch of the trigeminal nerve *(Moore, p 863).*

34. **(E)** The zygomatic branch of CN VII innervates the inferior part of the orbicularis oculi *(Moore, p 863).*

35. **(D)** The stylomastoid foramen is located between the styloid and mastoid processes of the temporal bone. CN VII and the stylomastoid branch of the posterior auricular artery travel through it *(Moore, p 862).*

36. **(E)** The facial artery has the following branches: inferior labial, superior labial, lateral nasal, and angular *(Moore, p 866).*

37. **(A)** The supratrochlear and supraorbital arteries are branches of the internal carotid artery, while most other arteries of the face are branches of the external carotid artery *(Moore, p 866).*

38. **(C)** The superficial temporal artery supplies the facial muscles and skin of the frontal and temporal regions, while the transverse facial artery supplies the parotid gland and duct as well as muscles and skin of the face *(Moore, p 866).*

39. **(B)** Within the parotid gland are found the facial nerve and its branches, the retromandibular vein, the external carotid artery, and parotid lymph nodes *(Moore, p 870).*

40. **(A)** The parasympathetic component of the glossopharyngeal nerve travels to the parotid gland via the auriculotemporal nerve *(Moore, p 870).*

41. **(A)** Parasympathetic fibers from CN IX synapse in the otic ganglion before traveling to the parotid gland *(Moore, p 870).*

42. **(D)** The scalp is composed of skin, connective tissue, aponeurosis, loose connective tissue, and pericranium *(Moore, p 872).*

43. **(E)** The cavernous sinus is situated on each side of the sella turcica. It extends on each side from the superior orbital fissure to the petrous part of the temporal bone *(Moore, p 880).*

44. **(A)** The cavernous sinus receives blood from the superior ophthalmic veins, the inferior ophthalmic veins, the superficial middle cerebral vein, and the sphenoparietal sinus *(Moore, p 880).*

45. **(A)** The cavernous sinus contains the internal carotid artery with its small branches, the carotid plexus of sympathetic nerves, the oculomotor nerve, the trochlear nerve, the trigeminal nerve (the ophthalmic division and occasionally the maxillary division), and the abducent nerve *(Moore, p 882).*

46. **(D)** The dura mater is innervated by all three divisions of the trigeminal nerve, C1–C3, and possibly the vagus nerve *(Moore, p 883).*

47. **(D)** The cavity of the medulla oblongata forms the inferior part of the fourth ventricle.

48. **(B)** Cerebrospinal fluid flows from each lateral ventricle through an interventricular foramen into the 3rd ventricle, then through the cerebral aqueduct into the 4th ventricle, and finally through median and lateral apertures into the subarachnoid space *(Moore, p 899).*

49. **(D)** The carotid canal is located in the inferior surface of the petrous temporal bone in the middle cranial fossa and contains the internal carotid artery, sympathetic plexus, and internal carotid venous plexus *(Moore, p 893).*

50. **(C)** The foramen cecum is unpaired *(Moore, p 846).*

51. **(D)** The oculomotor nerve emerges between the superior cerebellar artery and the posterior cerebral artery *(Moore, p 894).*

52. **(D)** The abducent nerve emerges between the labyrinthine artery and the anterior inferior cerebellar artery *(Moore, p 894).*

53. **(E)** The vertebral artery system has the following branches: posterior communicating, posterior cerebral, basilar, superior cerebellar, anterior inferior cerebellar, posterior inferior cerebellar, labyrinthine, and anterior spinal. The ophthalmic artery is a branch of the internal carotid artery system *(Moore, p 895).*

54. **(C)** The middle meningeal artery supplies more blood to the calvaria than to the dura, which is supplied by the vertebral artery *(Moore, pp 883, 895).*

55. **(A)** Tears flow from the lacrimal ducts across the eye to the lacrimal lake. The tears then pass through the lacrimal punctum on the lacrimal papilla, into the lacrimal canaliculi, and into the lacrimal sac, which is the widened end of the nasolacrimal duct *(Moore, pp 901–902).*

56. **(A)** Parasympathetic fibers from CN VII travel first via the greater petrosal nerve and then via the nerve of the pterygoid canal to the pterygopalatine ganglion, where they synapse with postsynaptic fibers. These fibers join the zygomatic branch of V_2, which in turn conveys the fibers to the lacrimal branch of V_1 (via a communicating branch). This nerve delivers the parasympathetic fibers to the lacrimal gland *(Moore, p 903).*

57. **(E)** The superior oblique takes its origin from the body of the sphenoid bone, while all four rectus muscles take origin from the common tendinous ring *(Moore, p 910).*

58. **(B)** The lateral rectus is innervated by the abducent nerve *(Moore, p 910).*

59. **(D)** The superior oblique abducts, depresses, and rotates the eyeball medially *(Moore, p 910).*

60. **(A)** The long ciliary nerves convey postsynaptic sympathetic fibers to the dilator pupillae and afferent fibers from the iris and cornea *(Moore, pp 911–912).*

61. **(B)** Afferent fibers from the iris and cornea pass through the ciliary ganglion *(Moore, p 912).*

62. **(E)** The infraorbital artery is a branch of the third part of the maxillary artery *(Moore, p 913).*

63. **(C)** The short posterior ciliary arteries pierce the sclera near the optic nerve and supply the choroid, which subsequently supplies the rods and cones of the retina *(Moore, p 913).*

64. **(E)** The infratemporal fossa contains the otic—not the pterygopalatine—ganglion *(Moore, p 920).*

65. **(A)** The buccinator is a muscle of facial expression, not a muscle of mastication *(Moore, p 921).*

66. **(A)** The muscles of mastication are associated with the first branchial arch and are innervated by the trigeminal nerve *(Moore, p 921).*

67. **(A)** The lateral pterygoids, when acting together, depress and protrude the mandible, with the assistance of the suprahyoid muscles, infrahyoid muscles, and gravity *(Moore, pp 921, 926).*

68. **(E)** The descending palatine artery is a branch of the third (pterygopalatine) part of the maxillary artery *(Moore, pp 920, 922).*

69. **(B)** The labyrinthine artery is a branch of the basilar artery, which is part of the vertebral artery system *(Moore, p 895).*

70. **(D)** The artery of the pterygoid canal supplies the superior part of the pharynx, the auditory tube, and the tympanic cavity *(Moore, p 922).*

71. **(E)** The sphenopalatine artery is transmitted through the sphenopalatine foramen along with the nasopalatine and superior nasal nerves *(Moore, pp 922, 950).*

72. **(C)** Postsynaptic parasympathetic fibers pass from the ganglion to the parotid gland via the auriculotemporal nerve *(Moore, pp 922–923).*

73. **(E)** The tensor tympani is innervated by the mandibular division of the trigeminal nerve *(Moore, p 971).*

74. **(C)** The temporomandibular joint is a modified hinge-type synovial joint *(Moore, p 923).*

75. **(A)** The hard palate is composed of the palatine processes of the maxillary bones as well as the horizontal plates of the palatine bones *(Moore, p 935).*

76. **(A)** The tensor veli palatini is innervated by the medial pterygoid nerve, which is a branch of the mandibular division of the trigeminal nerve *(Moore, p 939).*

77. **(E)** The tensor veli palatini tenses the soft palate and opens the mouth of the auditory tube during swallowing and yawning *(Moore, p 939).*

78. **(C)** The four types of lingual papillae are the vallate papillae, foliate papillae, filiform papillae, and fungiform papillae *(Moore, pp 940–941).*

79. **(D)** The hypoglossal nerve innervates the four intrinsic muscles of the tongue in addition to the following three extrinsic muscles of the tongue: genioglossus, hyoglossus, and styloglossus. The vagus innervates the palatoglossus, which is both an extrinsic tongue muscle and a palate muscle *(Moore, p 942).*

80. **(E)** The genioglossus depresses the tongue and assists in protrusion. The hyoglossus depresses and retracts the tongue. The styloglossus retracts the tongue and draws it up for swallowing. The palatoglossus elevates the posterior part of the tongue. The superior and inferior longitudinal muscles curl the tip and sides of the tongue and shorten it. The transverse narrows and elongates the tongue, assisting in protrusion. The vertical flattens and broadens the tongue, assisting in protrusion *(Moore, p 942).*

81. **(B)** In respect to general sensation (touch and temperature), the mucosa of the anterior two-thirds of the tongue is innervated by the lingual nerve, which is a branch of the mandibular division of the trigeminal nerve. The mucosa of the posterior one-third of the tongue is innervated by the lingual branch of the glossopharyngeal nerve, and the mucosa of the area just anterior to the epiglottis is innervated by small branches of the internal laryngeal nerve, which is a branch of the vagus. In respect to special sensation (taste), the chorda tympani, which is a branch of the facial nerve, innervates the anterior two-thirds of the tongue except for the vallate papillae in the posterior one-third of the tongue, which are innervated by the lingual branch of the glossopharyngeal nerve. Branches of the internal laryngeal nerve of the vagus innervate the area anterior to the epiglottis in respect to taste *(Moore, p 944).*

82. **(A)** Parasympathetic and taste fibers from the chorda tympani reach their destination by joining the lingual nerve, a branch of the mandibular division of the trigeminal nerve *(Moore, p 944).*

83. **(D)** Sweetness is detected on the apex, or tip, of the tongue, saltiness on the lateral margins, and sourness and bitterness on the posterior part *(Moore, p 944).*

84. **(A)** The dorsal lingual arteries supply the posterior tongue and the palatine tonsil, while the deep lingual artery supplies the anterior tongue. The sublingual artery supplies the sublingual gland and the floor of the mouth. The dorsal lingual veins accompany the lingual artery, and the deep lingual veins drain the apex of the tongue, ultimately joining the sublingual vein *(Moore, p 944).*

85. **(C)** The parotid gland is supplied by branches of the external carotid and superficial temporal arteries and innervated by the glossopharyngeal nerve. The submandibular gland is supplied by the submental artery and innervated by the parasympathetic fibers of the facial nerve that synapsed in the submandibular ganglion. The sublingual glands are supplied by the sublingual and submental arteries and innervated by parasympathetic fibers of the facial nerve *(Moore, p 948).*

86. **(E)** Openings to the pterygopalatine fossa include the pterygomaxillary fissure, sphenopalatine foramen, inferior orbital fissure, foramen rotundum, and pterygoid canal *(Moore, p 950).*

87. **(E)** The pterygopalatine fossa contains the third part of maxillary artery, the maxillary nerve, the nerve of the pterygoid canal, and the pterygopalatine ganglion *(Moore, p 950).*

88. **(A)** The inferior orbital fissure transmits the infraorbital and zygomatic branches of the maxillary nerve, the infraorbital vessels, the inferior ophthalmic veins, and the orbital branches of the pterygopalatine ganglion *(Moore, p 951).*

89. **(B)** The greater petrosal nerve (parasympathetic fibers of the facial nerve) joins the deep petrosal nerve (sympathetic nerves from the internal carotid plexus) to form the nerve of the pterygoid canal *(Moore, p 951).*

90. **(E)** The nerve of the pterygoid canal innervates the lacrimal gland, the palatine glands, the mucosal glands of the nasal cavity, and the mucosal glands of upper pharynx *(Moore, p 951).*

91. **(A)** The superior meatus is a narrow passage, inferior to the superior nasal concha, in which the posterior ethmoidal sinuses open. The middle meatus contains openings to the frontal sinus, maxillary sinus, and anterior and middle ethmoidal sinuses *(Moore, p 956).*

92. **(C)** The nasolacrimal duct drains into the inferior meatus *(Moore, p 956).*

93. **(C)** The inferior nasal concha is a separate bone, whereas the superior and middle nasal conchae are simply extensions of the ethmoid bone *(Moore, p 832).*

94. **(D)** The medial and lateral walls of the nasal cavity are supplied by the sphenopalatine artery, anterior and posterior ethmoidal arteries,

greater palatine artery, and superior labial artery *(Moore, p 956).*

95. **(A)** The nasal mucosa is innervated by the nasopalatine nerve, greater palatine nerve, anterior ethmoidal nerve, and posterior ethmoidal nerve *(Moore, p 956).*

96. **(E)** The auricle is innervated by the great auricular nerve and auriculotemporal nerve. The external surface of the tympanic membrane is innervated by the auriculotemporal nerve and even a small branch of the vagus. The internal surface of the tympanic membrane is innervated by the glossopharyngeal nerve. The pharyngotympanic tube is innervated by the tympanic plexus (fibers from the facial and glossopharyngeal nerves). The sensory cell bodies of the vestibulocochlear nerve are located in the spiral and vestibular ganglia *(Moore, pp 962, 967, 969, 975).*

97. **(B)** The dura mater of the floor of the middle cranial fossa is superior to the tegmental roof. The internal jugular vein is inferior to the floor. The lateral membranous wall is formed by the epitympanic recess and contains the head of the malleus. The medial labyrinthine wall separates the tympanic cavity from the internal ear, which contains the cochlea. The anterior carotid wall separates the tympanic cavity from the carotid canal. The posterior mastoid wall guards the mastoid cells and the facial nerve *(Moore, p 967).*

98. **(D)** The tympanic cavity contains the auditory ossicles, tympanic plexus, chorda tympani nerve, and stapedius and tensor tympani muscles *(Moore, p 967).*

99. **(A)** The pharyngotympanic tube opens posterior to the inferior meatus of the nasal cavity and serves to equalize the pressure in the middle ear with atmospheric pressure. The tensor veli palatini and levator veli palatini work together to open the tube, which is supplied by the ascending pharyngeal artery, middle meningeal artery, and artery of the pterygoid canal *(Moore, p 969).*

100. **(B)** The mandibular division of the trigeminal nerve innervates the tensor tympani *(Moore, p 971).*

101. **(C)** The tensor tympani assists in preventing damage to the internal ear when one hears loud noises by pulling the handle of the malleus medially, thereby tensing the tympanic membrane and reducing the amplitude of its oscillations *(Moore, p 971).*

102. **(C)** The internal acoustic meatus transmits the facial nerve, vestibulocochlear nerve, and labyrinthine artery *(Moore, p 976).*

103. **(E)** The saccule is continuous with the cochlear duct through the ductus reuniens *(Moore, p 975).*

104. **(C)** Typical cervical vertebrae (C3–C6) have transverse processes with foramina transversaria, which transmit both the vertebral vein and artery. In C7, however, this foramen transmits only the vertebral vein *(Moore, p 996).*

105. **(C)** The pretracheal layer of deep cervical fascia encloses, or invests, the infrahyoid muscles *(Moore, p 998).*

106. **(A)** The superior attachments of the trapezius are the medial third of the superior nuchal line, external occipital protuberance, ligamentum nuchae, spinous processes of C7–T12, and lumbar and sacral spinous processes *(Moore, p 1002).*

107. **(D)** The posterior triangle of the neck contains an anterior boundary formed by the posterior border of the SCM, a posterior boundary formed by the anterior border of the trapezius, an inferior boundary formed by the middle third of the clavicle, an apex where the SCM and trapezius meet on the superior nuchal line, a roof formed by the investing layer of deep cervical fascia, and a floor formed by the muscles covered by the prevertebral layer of deep cervical fascia *(Moore, p 1003).*

108. **(A)** The anterior triangle contains the submandibular (digastric), submental, carotid, and muscular (omotracheal) triangles. The poste-

rior triangle contains the occipital and supraclavicular (omoclavicular or subclavian) triangles *(Moore, p 1004).*

109. **(E)** The posterior cervical triangle contains the splenius capitis, levator scapulae, middle scalene, and posterior scalene *(Moore, p 1004).*

110. **(A)** The suprahyoid muscles are the mylohyoid, geniohyoid, stylohyoid, and digastric muscles *(Moore, p 1016).*

111. **(A)** The mylohyoid and anterior belly of the digastric are innervated by the trigeminal nerve *(Moore, p 1016).*

112. **(C)** The anterior vertebral muscles are the longus colli, longus capitis, rectus capitis anterior, and rectus capitis lateralis *(Moore, p 1026).*

113. **(A)** The only lateral vertebral muscle that is innervated by dorsal rami is the splenius capitis *(Moore, p 1026).*

114. **(A)** There is NOT a middle thyroid artery, but there is a middle thyroid vein *(Moore, pp 1030–1033).*

115. **(B)** Arytenoids, corniculate, and cuneiform cartilages are paired, whereas the thyroid, cricoid, and epiglottic are single *(Moore, p 1038).*

116. **(C)** All intrinsic laryngeal muscles are innervated by the recurrent laryngeal nerve except the cricothyroid, which is innervated by the external laryngeal nerve *(Moore, p 1045).*

117. **(A)** The cricothyroid stretches and tenses the vocal fold *(Moore, p 1045).*

118. **(E)** The posterior cricoarytenoid abducts the vocal fold *(Moore, p 1045).*

119. **(C)** The internal laryngeal nerve is the sensory nerve of the larynx *(Moore, p 1048).*

120. **(A)** The pharyngeal tonsils are commonly referred to as the adenoids *(Moore, p 1051).*

121. **(E)** The salpingopharyngeus is innervated by the glossopharyngeal nerve. The three constrictors, palatopharyngeus, and salpingopharyngeus are innervated by the cranial root of the accessory nerve *(Moore, p 1056).*

122. **(A)** The superior constrictor inserts on the median raphe of the pharynx and the pharyngeal tubercle on the basilar part of the occipital bone *(Moore, p 1056).*

123. **(B)** The ascending palatine artery, levator veli palatini, and pharyngotympanic tube pass through the gap between the superior constrictor and the skull *(Moore, pp 1055, 1058).*

124. **(A)** The glossopharyngeal nerve, stylopharyngeus, and stylohyoid ligament pass through the gap between the superior and middle constrictors *(Moore, p 1058).*

125. **(A)** The superior laryngeal artery and vein and the internal laryngeal nerve pass through the gap between the middle and inferior constrictors *(Moore, p 1058).*

126. **(D)** The inferior laryngeal artery and recurrent laryngeal nerve pass through the gap inferior to the inferior constrictor *(Moore, p 1058).*

127. **(A)** A Le Fort I fracture is a horizontal fracture of the maxillae *(Moore, p 836).*

128. **(B)** The mastoid processes are absent at birth. Therefore, the facial nerves are close to the surface when they emerge from the stylomastoid foramina and may be injured during delivery *(Moore, p 847).*

129. **(D)** The site of anesthetic injection to block the inferior alveolar nerve is the mandibular foramen *(Moore, p 861).*

130. **(E)** A lesion of the zygomatic branch of CN VII would cause paralysis of the orbicularis oculi and therefore a drooping of the lower eyelid. Subsequently, tears would fail to spread over the cornea, resulting in a corneal scar and therefore impaired vision *(Moore, p 864).*

131. **(C)** The facial veins make clinically important connections with the cavernous sinus through the superior ophthalmic veins. Infections in the orbit, nasal sinuses, and superior part of the face may lead to cavernous sinus thrombosis *(Moore, p 883).*

132. **(A)** An epidural hematoma consists of blood from the middle meningeal artery *(Moore, p 886).*

133. **(E)** Cerebral compression is typically attributed to intracranial collections of blood, obstruction of CSF flow, intracranial tumors or abscesses, and edema of the brain *(Moore, p 888).*

134. **(B)** Ptosis (drooping upper eyelid) is caused by a lesion to the oculomotor nerve, which innervates the levator palpebrae superioris *(Moore, p 903).*

135. **(E)** Interruption of the cervical sympathetic trunk results in Horner syndrome *(Moore, p 912).*

136. **(E)** Symptoms of Horner syndrome include pupillary constriction, ptosis, sinking in of one eye, and absence of sweating on the face and neck *(Moore, p 1030).*

137. **(C)** A lesion to the hypoglossal nerve (due to a neck laceration or basal skull fracture) would result in the protruded tongue deviating toward the affected side in addition to altered articulation *(Moore, p 1098).*

138. **(B)** Fracture of the cribriform plate might cause a lesion to the olfactory tract, resulting in anosmia or CSF rhinorrhea *(Moore, p 1098).*

139. **(C)** Sagging of the soft palate, deviation of the uvula to the normal side, and hoarseness might be caused by a lesion to the vagus nerve at the brainstem or in the neck *(Moore, p 1098).*

140. **(A)** A neck laceration might damage the spinal root of the accessory nerve, resulting in paralysis of the SCM and superior fibers of the trapezius and drooping of the shoulder *(Moore, p 1098).*

141. **(D)** Pressure from the uncus, a fracture in the cavernous sinus, or aneurysms could damage CN III, resulting in a dilated pupil, ptosis, the eye being turned down and out, and a loss of pupillary reflex on the side of the lesion *(Moore, p 1098).*

142. **(C)** A laceration or contusion in the parotid region, a fracture of the temporal bone, or a stroke might damage the facial nerve, resulting in paralysis of facial muscles, an eye that remains open, a drooping mouth, a smooth-appearing forehead (no wrinkles), dry cornea, and loss of taste in the anterior two-thirds of the tongue *(Moore, p 1098).*

143. **(C)** A fracture involving the cavernous sinus might damage the oculomotor nerve or the abducens nerve *(Moore, p 1098).*

144. **(C)** The third arch is associated with the glossopharyngeal nerve, stylopharyngeus, and the greater horn and lower portion of the body of the hyoid bone *(Sadler, p 348).*

145. **(A)** The arytenoid and cricoid cartilages and laryngeal connective tissue are formed by lateral plate mesoderm *(Sadler, p 345).*

146. **(E)** The lower lip is formed from the mandibular prominence *(Sadler, p 370).*

147. temporalis

148. masseter

149. buccinator

150. depressor anguli oris

151. depressor labii inferioris

152. auriculotemporal nerve

153. buccal nerve

154. lingual nerve

155. inferior alveolar nerve

156. mylohyoid nerve

157. palatine glands

158. greater palatine nerve

159. greater palatine artery

160. palatopharyngeal arch

161. uvula

162. medial rectus

163. inferior rectus

164. superior rectus

165. superior oblique

166. lateral rectus

167. inner hair cell

168. tectorial membrane

169. vestibular membrane

170. outer hair cell

171. basal membrane

172. trigeminal ganglion

173. maxillary nerve

174. anterior superior alveolar nerve

175. pterygopalatine ganglion

176. posterior superior alveolar nerve

177. petrosquamous suture

178. external acoustic meatus

179. mandibular fossa

180. mastoid process

181. styloid process

182. head of malleus

183. vestibular nerve

184. tympanic membrane

185. stapes

186. tensor tympani

References

Moore KL, Dalley AF: *Clinically Oriented Anatomy,* 4th ed. New York, Lippincott Williams and Wilkins, 1999.

Sadler TW: *Langman's Medical Embryology,* 8th ed. New York, Lippincott Williams and Wilkins, 2000.

Index

A
Abdominal muscles
internal, 49, 61
transverse, 47, 60
Abdominal oblique aponeurosis
external, 59, 66
internal, 59, 66
Abdominal oblique muscle, external, 47, 48, 60
"Abdominal policeman," 62
Abdominal wall, anterolateral, 47, 60
Abducens nerve, 125, 139
Abducent nerve, 115, 134
Abductor digiti minimi brevis muscle, 93, 104
Abductor digiti minimi muscle, 98, 107
Abductor hallucis muscle, 93, 104
Abductor pollicis longus muscle, 16, 25, 26
Accessory hemiazygous vein, 36, 43
Accessory nerve, cranial root of, 123, 138
Acetabular notch, 87, 100
Acetabulum, 87, 100
Acetylcholine, 8
Acoustic meatus
external, 130, 140
internal, 121, 137
Acromioclavicular joint, 18, 27
Acromion, fractures of, 18, 27
Adductor canal, 90, 102
Adductor hiatus, 89, 101
Adductor magnus muscle, 89, 101
Adductor pollicis muscle, 17, 26
Adductor tubercle, 87, 100
Adenoids, 123, 138
Alveolar nerve
inferior, 124, 127, 138, 139
superior, 130, 140
Alveoli, 32, 42
Ampullae
of rectum, support of, 74, 83
of uterine tubes, 70, 81
Anal canal, 71, 82
Anal columns, 71, 82
Anal sphincter, 77, 86
Anatomical snuff box, 16, 26
Anconeus muscle, 14, 25
Anesthesia, epidural, caudal, 5, 9
Ankle joint, 96, 106
dorsiflexion of, 92–93, 96, 104, 106
grip of malleoli on trochlea and, 97, 106
injury of, 96, 106
Anococcygeal ligament, 74, 83
Antebrachial cutaneous nerve, medial, 21, 29
Anterior compartment, shin splints and, 98, 107
Anus, 74, 84
Aorta, 39, 45, 49, 54, 61, 65
abdominal, 74, 83
bifurcation of, 55, 64–65
Aortic arch, 32, 35, 36, 38, 42, 43, 44–45
Aortic hiatus of diaphragm, 38, 44, 58, 66
Appendicular artery, 52, 63
Appendix, 52, 63
Appendix testis, 72, 83
Arch of foot, longitudinal, medial, 96, 106
Arcuate ligaments, lateral, 54, 65
Arcuate line, 48, 60–61
Articular processes, cervical, 1, 7
Arytenoid cartilage, 125, 139
Atria, 33, 42
Auricle, left, 39, 45
Auricular artery, posterior, stylomastoid branch of, 112, 133
Auriculotemporal nerve, 111, 113, 127, 132, 134, 139
Auscultation, triangle of, 19, 27–28
Axilla, 13–14, 24
injury of, 20, 28
Axillary artery, 14, 25
Axillary muscle, 19, 28
Axillary nerve, 13, 24, 25
Azygous vein, 33, 35, 36, 42, 43
B
Bacilliform papillae, 118, 135
Basal membrane, 129, 140
Biceps, short head of, 92, 103
Biceps brachii muscle, 21, 25, 29
Biceps femoris muscle, long head of, 99, 107
Biceps femoris tendon, 97, 107
Biceps muscle, 20, 28
Bile duct, 51, 52–53, 62, 63
Bladder
apex of, 77, 86
fundus of, 78, 86
neck of, in females, 73, 83
separation from pubic bones in females, 75, 84
trigone of, 69, 80
uvula of, 69, 80
Brachial artery, 14, 21, 25, 29
Brachialis muscle, 14, 20, 22, 24, 28, 29
Brachial plexus, 13–14, 24
injury of, 19, 28
Brachiocephalic vein, right, 39, 45
Brachioradialis muscle, 15, 25
Brain, ventricles of, fourth, 114, 134
Branchial arches, 125, 139
first, 116
Breast
blood supply of, 34, 43
cancer of, 31, 41
Bronchi
left main, 32, 41
right main, 32, 41, 55, 65–66
Bronchopulmonary segments, 32, 34–35, 41–42, 43
Buccal nerve, 127, 139
Buccinator muscle, 110, 116, 126, 132, 139
Buck's fascia, 85
Bulbocavernosus muscle, 77, 86
Bulbourethral glands, 69, 80
C
Calcaneofibular joint, 96, 106

Calcaneonavicular ligament, plantar, 96, 106
Calcaneus, 88, 96, 100, 106
Calculi, urinary, 75, 84
Calvaria, 109, 132
Caput medusae, 53, 63
Cardiac plexus, 36, 44
Cardiac vein, anterior, 33, 42
Cardiovascular system. *See also* Heart
  derivation of, 38, 44
Carotid artery, internal, 110, 112–113, 132, 133
Carotid canal, 114, 134
Carpals, 12, 23
Carpal tunnel, 17
Carpometacarpal joint, of thumb, 18, 27
Carpus, 12, 23
Caudal epidural anesthesia, 5, 9
Caval opening, 58, 66
Cavernous sinus, 113, 134
  fractures involving, 125, 139
Celia, 54, 65
Celiac ganglion, 44, 47
Celiac plexus, parasympathetic root of, 55, 64
Celiac trunk, 57, 66
Cephalic vein, 21, 29
Cerebellar arteries
  inferior, anterior, 115, 134
  posterior, 114, 134
  superior, 114, 134
Cerebral compression, 124, 139
Cerebrospinal fluid, flow of, 114, 134
Cervical fascia, deep, 121, 137
Cervical spine, 1, 5, 7, 8
  lesions in, 15, 25
Cervical triangle, posterior, 122, 138
Cervix, uterine, 70, 81
Chest pain, 34, 42, 44, 47
Chordae tendineae, 33, 42
Chorda tympani nerve, 117, 118–119, 135, 136
Chromaffin cells, 55, 65
Chyle cistern, 55, 65
Ciliary arteries, posterior, short, 116
Ciliary ganglion, 116
Ciliary nerves, long, 116
Circular folds, of jejunum, 51, 62–63
Circumflex artery, 35, 43
  femoral, medial, 90, 102
  medial, 94, 104–105
Clavicle, 11, 18, 23, 27, 33, 42
"Clawhand," 19, 28
Cleft hand, 20, 28
Clitoris
  development of, 72–73, 83
  prepuce of, 78, 86
Cluneal nerves
  inferior, 78, 86, 91, 103
  superior, 91, 103
Coccyx, 67, 77, 79, 85
Cochlear duct, 121, 137
Collateral ligaments
  fibular, 94, 104–105
  tibial, 94–95, 105
Colon
  ascending, 52, 55, 63, 66
  sigmoid, 52, 63
Connective tissue, loose, in extradural space, 3, 9
Constrictor muscle, superior, 123, 138
Conus arteriosus, 32, 42
Copula, 40, 45
Coracobrachialis muscle, 20, 22, 28, 29
Cornea, afferent fibers from, 116
Coronary arteries, 38, 44–45, 46–47
  right, 32, 35, 42, 43
Coronary sinus, 33, 42
Corpora cavernosa, 72, 74, 75, 78, 82, 84, 85, 86
Corpus spongiosum, 74, 75, 84, 85
Costal cartilage, 40, 45
Costoclavicular artery, 34, 43
Cranial fossa, 109, 132
  anterior, 110, 132
Cranial nerves
  abducens (sixth), 125, 139
  accessory (eleventh), 123, 138
  facial (eleventh), 111, 132
  facial (seventh), 114, 124, 125, 134, 138, 139
  glossopharyngeal (ninth), 123, 138
  hypoglossal (twelfth), 118, 124–125, 135, 139
  oculomotor (third), 114, 115, 124, 134–135, 139
  optic (second), 113, 119, 134, 136
  trigeminal (fifth), 110, 122, 132, 138
  vagus (tenth), 39, 44, 45, 47, 50, 55, 61–62, 64, 70, 81, 125, 139
Cremaster fascia, 49, 61
Cremaster muscle, 49, 61
Cribriform plate
  foramina in, 110, 132
  fracture of, 125, 139
Cricoarytenoid muscle, posterior, 123, 138
Cricoid cartilage, 125, 139
Cricothyroid muscle, 123
Cruciate ligaments, 95, 105
Crural fascia, 88, 101
Cubital fossa, 15, 25
Cutaneous nerve
  antebrachial
    lateral, 15, 25
    medial, 21, 29
  of arm, posterior, 12, 23
  brachial
    lateral, 25
    medial, 25, 34, 43
  femoral, lateral, 49, 59, 61, 66
Cystic artery, 53, 63
Cystic duct, 52–53, 63

D

Deep arteries
  of arm, 14, 24
  of thigh, 90, 102
Deferential artery, 76, 85
Deltoid muscle, 12, 18, 23, 27
Depressor anguli oris muscle, 126, 139
Depressor labii inferioris muscle, 126, 139
Depressor septi muscle, 111, 132
Diaphragm, 35, 43, 44, 47, 54, 65
  aortic hiatus of, 38, 44, 58, 66
  central tendon of, 54, 65
  domes of, 44, 47
  esophageal hiatus of, 38, 44, 55, 58, 64, 66
  inferior vena cava and, 44, 47
  inspiration and, 54, 65
  pelvic, musculofascial, 67, 79
Diaphragmatic crus, 55, 64
  left, 54, 65
  right, 58, 66
Diaphragmatic hernias, 55, 64
Digestion, 55, 66
Dorsalis pedis artery, 97, 98, 107
Dorsal nerve root, 33, 42
Dorsal rami, 3, 8, 34, 43, 122, 138
Ductus deferens, 49, 61
Duodenojejunal junction, 51, 62
Duodenum, 51, 53, 62, 64
  C-shaped curve of, 52, 63
  digestion in, 55, 66
  suspensory muscle of, 51, 62
  3rd part of, 56, 65
Dura mater, innervation of, 114, 134

E

Ear, innervation of, 120, 137
Elbow joint, 18, 27
  innervation of, 16, 26
Emissary veins, 110, 132
Epididymis, 49, 61
Epidural anesthesia, caudal, 5, 9
Epidural hematomas, 124, 139
Epidural space, 3, 8
Epigastric artery
  inferior, 48, 59, 60, 66
  superior, 59, 66
Episiotomy, 74, 84

Epo-ophoron, 72, 82–83
Erector spinae muscles, 2, 7
Esophageal hiatus of diaphragm, 38, 44, 55, 58, 64, 66
Esophagus, 55, 62, 65–66
  compression of, 36, 43
Ethmoidal foramina, anterior, 110, 132
Ethmoidal sinuses, posterior, 120
Extensor carpi radialis longus tendon, 16, 25
Extensor digitorum brevis muscle, 98, 107
Extensor digitorum longus muscle, 92–93, 98, 104, 107
Extensor pollicis brevis muscle, 16, 25
Extensor retinaculum muscle, inferior, 93, 104
Extradural space, 3, 8
Extraperitoneal cavity, 49, 61
Eyelids, lower, drooping, 124, 138

F
Facet joints, thoracic, 34, 42
Facial artery, 112, 113, 133
Facial expression, muscles of, 111, 132
Facial nerve, 111, 114, 125, 132, 134, 139
Facial skeleton, 109, 132
Facial veins, 124, 138–139
Falciform ligament, 56, 57, 65, 66
Fallopian tubes. *See* Uterine tubes
Fascia lata, 88, 100–101
Fat, in extradural space, 3, 9
Femoral artery, 90, 102
Femoral canal, 90, 102
Femoral nerve, 59, 66, 89, 90, 101–102
Femoral sheath, 89, 101–102
Femoral triangle, 89, 101
Femur, 87, 100
  head of
    blood supply to, 90, 102
    ligament of, 94, 104
  length of, 87, 100
  neck of
    blood supply to, 90, 102
    fractures of, 87, 97, 100, 107
Fibula, 88, 93, 100, 104
  neck of, 97, 107
Fibularis brevis muscle, 93, 104
Fibularis longus tendon, 96, 106
Fibular nerve, common, 92, 97, 103–104, 107
Flexor carpi radialis muscle, 15, 25
Flexor digitorum brevis muscle, 98, 107
Flexor digitorum profundus muscle, 15, 25
Flexor hallucis longus muscle, 93, 104
Flexor pollicis brevis muscle, 17, 26
Flexor pollicis longus muscle, 15, 16, 25, 26
Foot
  dorsiflexion of, 93, 98, 104, 107
  eversion of, 93, 104
  inversion of, 93, 104
Foramen cecum, 114, 134
Foramen magnum, 109, 110, 132
Foramen ovale, 119, 136
Foramen rotundum, 110, 132
Foramen spinosum, 110, 132
Foramen transversarium, 121, 137
Foregut, ventral wall of, 38, 45
Fornix, vaginal, posterior, 70, 80–81
Fourchette, 74, 84
Fractures
  of acromion, 18, 27
  of femoral neck, 87, 97, 100, 107
  Le Fort I, 124, 138
  of scapula, 18, 27
  temporal, 125, 139
  tibial, 88, 100
Frenulum, of labia minora, 74, 84
Frontal bone, 109, 132
Frontal nerve, 112, 133
Frontal suture, 109, 132
Fundus
  of bladder, 78, 86
  of uterus, 77, 86

G
Gallbladder, neck of, 53, 63
Gastric artery, left, 57, 66
Gastrocnemius muscle, 93, 104
Gastrocolic ligament, 57, 66
Gastroduodenal artery, 51, 62
Gastrohepatic ligament, 62
Gastrolienal ligament, 57, 66
Gastro-omental artery, left, 50–51, 56, 62, 65
Geniculate ganglion, 120, 137
Genital swellings, 72, 83
Genital tubercle, 72–73, 83
  in male, 72, 83
Genitofemoral nerve, 59, 66
  genital branch of, 49, 61
Genu valgum, 97, 106
Glans penis, 71, 74, 75, 82, 84–85
Glenohumeral joint, 17, 18, 26, 27
Glenoid cavity, 11, 23
Glossopharyngeal nerve, 123, 138
Gluteal artery, superior, 75, 84
Gluteal nerve, superior, 68, 79, 91, 103
Gluteus maximus muscle, 90–91, 102
Gluteus medius muscle, 91, 102
Gluteus minimus muscle, 91, 97, 99, 102, 107
Gracilis muscle, 89, 101
Great toe, lateral deviation of, 96, 106
Gubernaculum, 49, 61
Gynecoid pelvis, 73, 83

H
Hair cells
  inner, 129, 140
  outer, 129, 140
Hallux valgus, 96, 106
Hamstring muscles, 91, 92, 101, 103
Hand, innervation of, 16–17, 26
Heart
  atria of, 32, 33, 42
  interventricular septum of, 32, 42
  location of, 32, 42
  pain and, 44, 47
  sympathetic stimulation in, 44, 46–47
  valves of, 33, 42
  ventricles of, 33, 42
Helicine arteries, 74, 84
Hemiazygous vein, 35, 36, 43
Hepatic artery
  common, 53, 57, 63, 66
  proper, 57, 66
Hepatic veins, 53, 63
Hepatoduodenal ligament, 53, 63
Hernias
  diaphragmatic, 55, 64
  inguinal, direct, 48, 61
Hinge joints, 18, 27, 117, 135
Hip joint
  blood supply to, 94, 104–105
  flexion of, 97, 107
  fractures of, 87, 100
  hyperextension of, 94, 104
  rotation of, 97, 107
Hoarseness, 125, 139
Horner syndrome, 124, 139
Humerus, 18–19, 27
  condyle of, 11, 23
Humpback (hunchback), 5, 8
Hypogastric nerve, pelvic, 68, 79
Hypogastric plexus, inferior, 68, 79
Hypoglossal nerve, 118, 124–125, 135, 139
Hypomere, 20, 29
Hypothenar eminence, 17, 26

I
Ileocecal fold, inferior, 56, 65
Ileocolic artery, 52, 63
Ileum, innervation of, 51, 62
Iliac artery, internal, 68–69, 76, 80, 85

Iliac spines, superior, anterior, 67, 79
Iliofemoral ligament, 94, 104
Iliohypogastric nerve, 59, 66
Ilioinguinal nerve, 48, 61
Iliopsoas muscle, 88, 97, 101, 107
Iliopubic tract, 49, 61
Ilium, abdominal aorta and, 55, 64–75
Incisive foramen, 117, 135
Inferior vena cava, 39, 44, 45, 47, 53, 55, 63, 65
Infraorbital artery, 116, 120, 136–137
Infraspinatus muscle, 12, 23
Infratemporal fossa, 116
Infratrochlear nerve, 112, 133
Inguinal canal, 48, 61
Inguinal fossae, medial, 48, 61
Inguinal hernias, direct, 48, 61
Inguinal ligament, 47–49, 60, 61
Inguinal rings, 48, 61
  deep, 48, 60
Inion, 109, 132
Inspiration, diaphragm and, 54, 65
Intercarpal joints, 18, 27
Intercostal arteries
  anterior, 34, 43
  posterior, 34, 43
Intercostal muscles, 34, 43
Intercostal veins, posterior, 36, 43
Intercostobrachial nerve, 12, 23, 34, 43
Interosseous muscles, dorsal, 17, 26
Interosseous nerve, anterior, 16, 26
Interventricular septum, 32, 42
Intervertebral discs, cervical, 1, 7
Iris, afferent fibers from, 116
Ischial bursa, 91, 102
Ischial tuberosities, 68, 70–71, 77, 79, 81, 85, 91, 92, 102, 103
Ischiocavernosus muscle, 71, 72, 77, 81, 82, 86

J
Jejunum
  circular folds of, 51, 62–63
  innervation of, 51, 62
Jugular foramen, 110, 132
Jugular vein, internal, left, 39, 45

K
Kidneys. *See also* Renal *entries*
  diaphragmatic hernias and, 55, 64
  innervation of, 54, 65
  posterior surfaces of, 53, 64
Knee joint, 89, 101
  flexion of, 93, 104
  ligaments of, 95, 105
  menisci of. *See* Menisci
  muscle stabilizing, 94, 104–105
  stabilization of, 97, 106–107
  "unhappy triad" of injuries to, 95, 105
Kyphosis, 5, 8

L
Labia minora, 72, 82
  development of, 73, 83
  frenulum of, 74, 84
Labyrinthine artery, 115, 117, 134, 135
Lacrimal gland, parasympathetic fibers to, 115, 134
Lacrimal nerve, 111, 112, 132, 133
Lacrimation, 124
Lacunar ligament, 49, 61
Lambda, 109, 132
Lambdoid suture, 109, 132
Laryngeal artery
  inferior, 123, 138
  superior, 123, 138
Laryngeal cartilages, 122, 138
Laryngeal muscles, 122–123, 138
Laryngeal nerve, internal, 123, 138
Larynx
  connective tissue of, 125, 139
  sensory muscle of, 123, 138
Lateral cord, 21, 29
Lateral plate mesoderm, 125, 139
Latissimus dorsi muscle, 19, 28
Le Fort I fractures, 124, 138
Leptomeningeal space, 4, 8
Levator ani muscle, 74, 76, 77, 83, 85, 86
Levatores costarum, 34, 43
Levator scapulae muscles, 13, 24
Ligament of Treitz, 51, 62
Ligamentum arteriosum, 39, 45
Limbs, innervation of, 3, 8
Linea alba, 47, 60
Lingual arteries, dorsal, 119, 136
Lingual nerve, 118–119, 127, 136, 139
Lingual papillae, 118, 135
Linguals, 32, 39, 41, 45
  of left lung, 34–35, 43
Lips, elevation and protrusion of, 111, 132
Lithotripsy, 75, 84
Liver
  caudate lobe of, 62
  lobes of, 56, 65
  round ligament of, 53, 63
Lobster claw deformity, 20, 28
Longissimus muscle, 2, 7
Longus capitis muscle, 3, 8
Longus colli muscle, 2, 7
Lordosis, during pregnancy, 8
Lumbar spine, 1, 5, 7, 9
Lumbricals, 17, 98, 107
Lunate surface, of acetabulum, 87, 100
Lungs
  development of, 38, 45
  gas exchange in, 32, 42
  left, bronchopulmonary segments of, 34–35, 43
  right, 32, 41
  surfactant in, 44, 47
  sympathetic stimulation of, 44, 47
Lymphatic system, breast cancer and, 31, 41

M
Malleoli
  grip on trochlea, 97, 106
  lateral, 87, 100
  medial, 87, 100
Malleus, head of, 131, 140
Mammary glands, 34, 43
Mandibles, depression of, 116–117, 135
Mandibular foramen, 124, 138
Mandibular fossa, 130, 140
Mandibular nerve, 121, 137
Mandibular prominence, 125, 139
Manubrium, of sternum, 40, 45
Masseter muscle, 111, 126, 132, 139
Mastication, muscles of, 110, 116, 132
Mastoid process, 124, 130, 138, 140
  lateral surface of, 122, 137
Maxillary artery, 117, 135
Maxillary bones, 118, 135
Maxillary nerve, 112, 130, 133, 140
Medial epicondyle, 19, 27
Median nerve, 15–17, 20, 21, 25, 26, 28, 29
  recurrent branch of, 17, 26
Mediastinum, 35, 43
Medulla oblongata, 114, 134
Meningeal artery, middle, 111, 115, 124, 132, 134, 139
Meninges
  arterial branches to, 110, 132
  nerve branches to, 110, 132
Meningomyelocele, 5, 9
Menisci, 95, 105
  lateral, 94, 95, 105
    pain related to, 96, 106
  medial, 94–95, 105
Mentalis muscle, 111, 132
Mental nerve, 112, 133
Mesenteric artery
  inferior, 74, 84
  superior, 56, 65
Mesenteric vein, superior, 56, 65
Mesentery, root of, 51, 62
Mesoderm, 38, 44

Mesometrium, 76, 85
Mesonephric system, 72, 82–83
Mesosalpinx, 73, 83
Metatarsal bones, second, 88, 100
Metopic suture, 109, 132
Midclavicular planes, 47, 60
Midgut, embryonic development of, 56, 65
Musculocutaneous nerve, 15, 21, 25, 29
Musculophrenic arteries, 34, 43
Mylohyoid muscle, 122, 138
Mylohyoid nerve, 127, 139
Myoblasts, 20, 29

N
Nasal bones, 120
Nasal cavity
  blood supply to, 120, 136–137
  nasolacrimal duct communication with, 120
Nasal conchi, inferior, 120
Nasal mucosa, innervation of, 120, 137
Nasal nerve, external, 111, 132
Nasolacrimal duct, 120
Navicular, 88, 100
Navicular fossa, 71, 75, 82, 84–85
Neck. *See also* Cervical spine
  posterior triangle of, 122, 137
  superficial lacerations of, 125, 139
Nelaton's line, 92, 103
Nerve supply
  of back, 3, 8
  of upper limb, 12, 13–14, 24
Neurocranium, 109, 132

O
Oblique aponeurosis, external, 48, 60
Oblique fissure, 40, 45
Oblique muscle
  abdominal, internal, 59, 66
  superior, 115, 116, 129, 134, 140
Obturator artery, 69, 80
Obturator externus tendon, 94, 104
Obturator fascia, 71, 81
Obturator internus muscle, 68, 76, 79, 85
Obturator nerve, 71, 75, 82, 85
Occipital bone, pharyngeal tubercle of, 123, 138
Occipital nerve, lesser, 3, 8
Occipital protuberance, external, 109, 132
Oculomotor nerve, 114, 115, 124, 134–135, 139
Olfactory cells, axons of, 110, 132
Olfactory tract, lesions to, 125, 139
Omental appendices, 52, 63
Omental foramen, 62
Omentum
  greater, 62
  lesser, 57, 62, 66
Ophthalmic arteries, 110, 115, 116, 132, 134
Ophthalmic nerve, 111, 112, 132, 133
Ophthalmic veins, superior, 124, 138–139
Optic nerve, 113, 119, 134, 136
Orbicularis oculi muscle, 111, 112, 132, 133
Orbital fissure
  inferior, 119, 136
  superior, 110, 132
Os, external, 70, 81
Osteoporosis, 5, 8
Otic ganglion, 113, 117, 134, 135
Ovarian arteries, 74, 83
Ovaries, suspensory ligament of, 70, 81

P
Palate
  hard, 118, 135
  muscles of, 118, 135
  soft, sagging of, 125, 139
Palatine artery
  ascending, 123, 138
  descending, 117, 135
  greater, 128, 140
Palatine bones, 109, 132
Palatine glands, 128, 140
Palatine nerve, greater, 128, 140
Palatoglossus muscle, 118, 135
Palatopharyngeal arch, 128, 140
Palmaris brevis muscle, 17, 26
Palmaris longus muscle, 16, 26
Palmaris longus tendon, 15, 25
Pampiniform plexus, 50, 61
Pancreas, 52, 54, 63, 65
  head of, 52, 63
Pancreatic duct, 51, 52–53, 62, 63
Pancreaticoduodenal arteries, 51, 62
Papillary muscle, posterior, 33, 42
Paramesonephric ducts, 72, 82–83
Paranasal sinuses, ethmoidal, 120
Parasympathetic fibers, 4, 8, 36, 44
  to bladder, 69, 80
  celiac plexus and, 55, 64
  from chorda tympani, 118–119, 136
  colonic, 52, 63
  inhibition of, 44, 46–47
  to lacrimal gland, 115, 134
  in otic ganglion, 117, 135
  to parotid gland, 113, 134
  of pelvic splanchnic nerves, 76, 85
  pelvic splanchnic nerves and, 55, 65
  postsynaptic, 4, 8, 117, 135
    in otic ganglion, 117, 135
  in pulmonary plexus, 44, 47
  testicular, 50, 61–62
Parasympathetic stimulation, engorgement of erectile tissue in bulbs of vestibule by, 72, 82
Paraumbilical vein, 53, 63
Paraurethral glands, 69, 75, 80, 84
Paraxial mesoderm, rib formation from, 38, 44
Parotid duct, 113, 133
Parotid gland, 113, 133
Parotid injuries, 125, 139
Patellar ligament, 95, 105
Pecten pubis, 67, 79
Pectinate line, 71, 82
Pectinate muscle, 33, 42
Pectoral arteries, 35, 43
Pectoralis major muscle, 18, 20, 27, 29
Pectoralis minor muscle, 12, 23
Pectoral nerve, lateral, 14, 21, 24, 29
Pelvic cavity, 67, 79
  formation of, 67, 79
Pelvic floor, 68, 79
Pelvic inlet, 67, 79
Pelvic wall, 68, 79
Pelvis
  gynecoid, 73, 83
  weak areas of, 68, 79
Penis, 72, 82. *See also* Glans penis
Pericardiacophrenic ligament, 33, 42
Pericardiacophrenic vessels, 39, 45
Pericardium, 33, 35, 42, 43
  fibrous, 38, 44
Perineal body, 71, 74, 81, 84
Perineal membrane, 75, 84
Perineal muscle, transverse, 77, 86
  deep, 75, 84
Perineal pouch
  deep
    in females, 71, 81
    muscles of, 71, 81–82
  superficial
    in males, 71, 81
    muscles of, 71, 81–82
Perineum, 74, 84, 90, 102
  boundaries of, 70, 81
  division into triangles, 70–71, 81
Peritoneal cavity, 62
Peritoneal fluid, 62
Peroneal nerve, deep, 98, 107
Petrosal nerves, 119, 136
  lesser, 120, 137
Petrosal sinuses, 113, 134
Petrosquamous suture, 130, 140
Phallus. *See* Glans penis; Penis

Pharyngeal arches, muscles of facial expression development from, 111, 132
Pharyngeal tonsils, 123, 138
Pharyngotympanic tube, 120–121, 137
Phrenic arteries, inferior, 35, 43, 54, 65
Phrenic nerve, 39, 45
Phrenic veins, superior, 36, 43
Piriformis muscle, 68, 76, 79, 85
Piriformis syndrome, 92, 103
Plantaris muscle, 93, 104
Plantaris tendon, grafting of, 97, 107
Plantar nerve
  lateral, 98, 107
  medial, 94, 104
Platysma, 122, 137
Pleura, parietal, 32, 41
Pleural cavity, 31, 41
Pleural fluid, 31, 41
Pleuropericardial membranes, 38, 44
Plicae circulares, of jejunum, 51, 62–63
Poland anomaly, 20, 29
Popliteal artery, 95, 105
Popliteal fossa, 92, 103
  floor of, 92, 103
Popliteal ligament, oblique, 92, 95, 103, 105
Popliteal vein, 88, 101
Popliteus muscle, 93, 94, 104–105
Porta hepatis, 53, 63
Portal triad, 62
Portal vein, 51, 62
Posterior longitudinal ligament, 2, 7
Pott's fracture-dislocation, 97, 106
Pregnancy, lordosis during, 8
Prepuce, of clitoris, 78, 86
Presynaptic fibers, 36, 44
Prevertebral ganglia, 36, 44
Prevertebral muscles, 35, 43
Primary rami
  dorsal, 91, 103
  ventral, 91, 103
Pronator quadratus muscle, 15, 25
Prostate, 73, 75, 78, 83, 84, 86
  lateral lobes of, 75, 84
Prostatic ductules, 69, 80
Prostatic sinus, 69, 80
Psoas, 55, 65
Pterygoid canal
  artery of, 117, 135
  nerve of, 119, 120, 136
Pterygoid muscle, lateral, 116–117, 135
Pterygopalatine fossa, 119, 136
Pterygopalatine ganglion, 112, 116, 130, 133, 140
Ptosis, 124, 139
Pubic bones, separation from bladder in females, 75, 84
Pubic ramus
  inferior, 67, 77, 79, 85
  superior, 67, 79
Pubic symphysis, 77, 85, 86
  anterior aspect of, 67, 79
Pubococcygeus muscle, 73, 83
Pubovesical ligaments, 73, 83
Pudendal artery, internal, 92, 103
Pudendal canal, 71, 81
Pudendal nerve, 68, 71, 76, 79, 81–82, 85, 90, 102
  perineal branches of, 78, 86
Pudendum, 74, 84
Pulmonary arteries, 39, 45
  right, 39, 45
Pulmonary plexus, 44, 47
Pulmonary veins, 32, 42
  inferior, left, 39, 45
  left, 39, 45
Pylorus, 57, 66

Q

Quadrangular space, 14, 24
Quadratus lumborum fasciae, 54, 65
Quadratus lumborum muscle, 58, 66
Quadriceps femoris muscle, 89, 94, 97, 101, 104–105, 106–107

R

Radial artery, 16, 26
Radial nerve, 12, 17, 21, 23, 25, 26, 29
Radius, 11, 23
Rami communicantes, 4, 8
Rectal artery, superior, 74, 84
Rectal nerve, inferior, 78, 86
Rectal venous plexus, internal, 71, 82
Rectal vessels, superior, 71, 82
Rectosigmoid junction, 70, 81
Rectouterine pouch, 77, 86
Rectovesical pouch, 78, 86
Rectovesical septum, 73, 83
Rectum, 52, 63
  ampulla of, support of, 74, 83
  male, anatomical relationships of, 74, 83
Rectus abdominis muscle, 48, 59, 60, 66
Rectus capitis posterior major muscle, 3, 8
Rectus femoris muscle, 89, 101
Rectus muscle
  inferior, 129, 140
  lateral, 115, 129, 134–135, 140
  medial, 129, 140
  superior, 129, 140
Rectus sheath, 48, 60–61
Renal calyces, minor, 53, 64
Renal hilum, 54, 65
Renal papillae, 53, 64
Renal pyramids, 54, 65
Renal vein, left, 52, 63
Respiratory system, 38, 45
Retromammary space, 31, 41
Retromandibular artery, 112, 133
Retropubic space, 75, 84
Rhomboid muscles, 12–13, 19, 24, 28
Ribs
  elevation of, 34, 43
  formation of, 38, 44
  transverse process articulation with, 31, 41
Rotator cuff injuries, 19, 28
Rotatores muscles, 2–3, 7
Round ligaments
  of liver, 53, 63
  of uterus, 73, 83
Rugae, 56, 62, 65

S

Saccule, 121, 137
Sacral spine, 70, 81
Sacrospinous ligament, 70, 81
Sacrotuberous ligament, 77, 85
Sagittal suture, 109, 132
Salivary glands, submandibular, 119, 120, 136
Saphenous nerve, 94, 104
Saphenous vein
  great, 88, 92, 101, 103
  small, 88, 101
Sartorius muscle, 88–89, 90, 101, 102
Scalp, layers of, 113, 134
Scaphoid, 12, 23
Scapula, 11, 23
  fractures of, 18, 27
  innervation of skin over, 14, 24
  muscles attached to coracoid process of, 12, 23
  "winging" of, 19, 27
Scapular nerve, dorsal, 12–13, 24
Sciatic foramen
  greater, 68, 79
  lesser, 75, 84, 90, 102
Sciatic nerve, 68, 79, 92, 99, 103, 107
  fibular division of, 92, 103
Sciatic notch, lesser, 67, 79
Scoliosis, 5, 8–9
Scrotal ligament, 49, 61
Scrotum
  development of, 72, 83
  innervation of, 49, 61, 71, 75, 82, 85
Sella turcica, 113, 134
Semimembranosus muscle, 95, 105
Seminal vesicles, 69, 73, 78, 80, 83, 86
Semitendinosus muscle, 92, 99, 103, 107

Serratus anterior muscle, 47, 60
Serratus posterior inferior muscle, 2, 7, 34, 43
Shin splints, 98, 107
Shoulder dislocations, 18, 27
Sigmoid colon, 52, 63
Sinoatrial node, 33, 42
Snuff box, 16, 26
Soleus muscle, 99, 107
Sphenopalatine artery, 117, 135
Sphenopalatine nerve, 120, 137
Spina bifida, 5, 9
Spinal arteries, 2, 4, 7, 8
Spinal cord
- enlargement for innervation of limbs, 3, 8
- veins of, 4, 8

Spinalis muscle, 2, 7
Spinal nerves, 4, 8, 33, 42
Spiral valve, 53, 63
Splanchnic nerves
- greater, 44, 47, 55, 64
- least, 44, 47
- lesser, 44, 47, 55, 64
- origin of, 44, 47
- pelvic, 52, 55, 63, 65, 68, 69, 79, 80
  - parasympathetic fibers in, 76, 85
- thoracic
  - lesser, 38, 44
  - lower, 44, 47

Spleen, 52, 54, 56, 63, 65
Splenic arteries, 50–51, 52, 56, 57, 62, 63, 65, 66
Splenius capitis muscle, 2, 7, 122, 138
Splenius cervicis muscle, 2, 7
Spondylolysis, 73, 83
Stapes, 131, 140
Sternal angle, 31, 41
Sternoclavicular joint, 17, 34, 43
Sternum, manubrium of, 40, 45
Stomach, 54, 65
- digestion in, 55, 66
- innervation of, 51, 62
- rugae of, 56, 62, 65

Styloid process, 130, 140
Stylomastoid branch foramen, 112, 133
Stylopharyngeus muscle, 123, 125, 139
Subarachnoid space, 4, 8
Subclavian artery, groove for, 31, 41
Subcostal nerve, 59, 66
Submandibular gland, 119, 120, 136
Suboccipital triangle, 3, 8
Subscapular artery, 13, 24
Superior oblique muscle, 3, 8
Superior vena cava, 39, 45
Supinator muscle, 16, 25
Supraclavicular nerves, 25
Supraclavicular triangle, 122, 137–138
Suprahyoid muscles, 121, 122, 137, 138
Suprarenal arteries, superior, 54, 65
Suprarenal cortex, 54, 65
Suprarenal glands, 54, 65
- innervation of, 54, 65
- left, 54, 65

Suprarenal medulla, 55, 65
Supraspinatus muscle, 13, 24
Supraspinatus tendon, 19, 28
Supratrochlear artery, 112–113, 133
Sural nerve, lateral, 92, 103–104
Surfactant, 44, 47
Suspensory ligament, of ovaries, 70, 81
Sustentaculum tali, 88, 100
Sweat glands, 34, 36, 43, 44
Sympathetic nerves, 35, 43
Sympathetic nervous system, 36, 44
- stimulation in heart and, 44, 46–47
- stimulation in lungs and, 44, 47

Sympathetic plexus, 110, 132
Sympathetic trunks
- cervical, 124, 139
- sacral, 68, 79

Syndactyly, 20, 28
Synovial joints, 18, 27, 117, 135

T

Talus, 87, 88, 100
Tarsal joint, transverse, 96, 106
Taste fibers, from chorda tympani, 118–119, 136
Taste sense, 118–119, 136
Tears, pathway followed by, 115, 134
Tectorial membrane, 129, 140
Temporal artery, superficial, 113, 133
Temporal bone, fracture of, 125, 139
Temporalis muscle, 111, 126, 132, 139
Temporomandibular joint, 117, 135
Tendinous intersections, of rectus abdominis muscle, 48, 60
Tendinous ring, common, 115, 134
Teniae coli, 52, 63
Tensor tympani, 121, 131, 137, 140
Tensor veli palatini, 118, 135
Testes, 49, 61
- development of, 49, 61
- thermoregulation and, 50, 61

Testicular artery, 68, 79
Thermoregulation, of testes, 50, 61
Thoracic artery, internal, 31, 35, 41, 43
Thoracic duct, 38, 44, 55, 64, 65
Thoracic fascia, anterior, 38, 44
Thoracic spine, 1, 5, 7, 9, 34, 36, 42, 43
Thoracoabdominal artery, 4, 8
Thymus, 35, 43
Thyroid artery, 122, 138
Thyroid disorders, 34, 42
Thyroid vein, inferior, 39, 45
Tibia, 88, 93, 94, 100, 104–105
- in genu valgum, 97, 106
- pain on lateral rotation on femur, 96, 106

Tibialis anterior muscle, 93, 98, 104, 107
Tibial nerve, 94, 104
- entrapment of, 96, 106

Tongue
- hypoglossal nerve lesions and, 124–125, 139
- innervation of, 118, 136
- muscles of, 118, 135

Tonsils, pharyngeal, 123, 138
Trabeculae carneae, 32, 42
Transumbilical plane, 47, 60
Transversalis fascia, 47, 48, 49, 59, 60, 61, 66
Transverse ligament, of knee, 95, 105
Transverse processes, 121, 137
Transversospinalis muscles, 2, 7
Transversus abdominis aponeurosis, 59, 66
Transversus abdominis muscle, 59, 66
Trapezius muscle, 11, 13, 23, 24, 137
Trendelenburg sign, 91, 102–103
Triceps brachii muscle, 14, 24
- lateral head of, 22, 29
- long head of, 22, 29

Triceps coxae, 91, 103
Triceps surae, 93, 104
Tricuspid valve, 33, 42
Trigeminal ganglion, 130, 140
Trigeminal nerve, 122, 138
- maxillary division of, 110, 132

Trigone, of bladder, 69, 80
Trochanter, greater, 92, 103
Trochlea, grip of malleoli on, 97, 106
Tunica albuginea, 49, 61
Tunica dartos, 75, 85
Tunica vaginalis, 49, 61
Tympanic cavity, 120, 137
Tympanic membrane, 131, 140

U

Ulna, 11, 22, 23, 29
Ulnar nerve, 15, 16–17, 19–21, 25–29
Umbilical arteries, 48, 61
Umbilical folds, 48, 61
Umbilical ligament, median, 73, 83
Umbilical vein, 53, 63
Uncus, herniating, 125, 139
"Unhappy triad" of knee injuries, 95, 105

Urachal remnants, 73, 83
Ureters, 69, 74, 76, 80, 83, 85
Urethra
  external orifice of, 69, 80
  male, 75, 84
  prostatic part of, 69, 80
Urethral folds, 73, 83
Urethral sphincter, external, 71, 75, 81, 84
Urinary bladder. *See* Bladder
Urinary calculi, 75, 84
Urogenital hiatus, 76, 85
Uterine artery, ligation of, 75, 84
Uterine tubes, 73, 77, 83, 86
  ampullae of, 70, 81
Uterus, 76, 85
  cervix of, 70, 81
  fundus of, 77, 86
  mesentery of, 76, 85
  position of, 70, 81
  round ligament of, 73, 83
Uvula, 128, 140
  of bladder, 69, 80
  deviation of, 125, 139

V
Vagina, 74, 76, 84, 85
  innervation of, 76, 85
  posterior fornix of, 70, 80–81
Vaginal fornices, 72, 83
Vagus nerve, 50, 61–62, 70, 81, 125, 139
  left, 39, 45
  posterior trunk of, 55, 64
Vasculature, of vertebral column, 2, 4, 7, 8
Vasoconstriction, 36, 44
Veins, of spinal cord, 4, 8
Ventricles
  of brain, fourth, 114, 134
  left, 33, 42
Vertebrae. *See also specific regions of spine*
  transverse processes of, 31, 41
Vertebral arch, in spondylolysis, 73, 83
Vertebral arteries, 115, 134
Vertebral column, 1, 7
Vertebral foramina, thoracic, 1, 7
Vertex, of neurocranium, 109, 132
Vesicle, inferior, 49, 61, 76, 85
Vestibulae
  bulb of, 78, 86
  female, 72, 74, 82, 84
Vestibular glands, lesser, 74, 84
Vestibular membrane, 129, 140
Vestibular nerve, 131, 140
Vestibulocochlear nerve, 120, 137
Visceral afferent fibers, 35, 43
Vocal folds, tensor of, 123, 138
Vulva, 74, 84

W
Water, reabsorption of, 55, 66
Wrist joint, abduction of hand at, 16, 26

X
Xiphoid process, 40, 45

The University of Bolton
The Library
Enquiries 01204 903094

Borrowed Items 01/02/2013 15:25
XXX6883

Item Title	Due Date
301428	
Modular deficits in Alzheim	08/02/2013
375288	
Anatomy & physiology for	08/02/2013
392004	
Appleton & Lange review c	08/02/2013

Amount Outstanding : £0.10

* = Newly borrowed items
Thankyou for using this unit

find us http://www.bolton.ac.uk/library

catalogue http://opac.bolton.ac.uk
PLEASE RETAIN YOUR RECEIPT